Geology of National Parks of Central/Southern Kenya and Northern Tanzania

Roger N. Scoon

Geology of National Parks of Central/Southern Kenya and Northern Tanzania

Geotourism of the Gregory Rift Valley,
Active Volcanism and Regional Plateaus

 Springer

Roger N. Scoon
Department of Geology
Rhodes University
Grahamstown
South Africa

Every effort has been made to contact the copyright holders of the figures and tables which have been reproduced from other sources. Anyone who has not been properly credited is requested to contact the publishers, so that due acknowledgment may be made in subsequent editions.

ISBN 978-3-030-08858-3   ISBN 978-3-319-73785-0   (eBook)
https://doi.org/10.1007/978-3-319-73785-0

Cover photographs: Front cover (and p. ii): The Ash Cone in the giant caldera of Mount Meru, Arusha National Park in northern Tanzania, which last erupted in 1910 (photograph by the author). Back cover: The author and his wife.

Printed on acid-free paper

This Springer imprint is published by the registered company Springer International Publishing AG part of Springer Nature
The registered company address is: Gewerbestrasse 11, 6330 Cham, Switzerland

The Western Escarpment of the Gregory Rift Valley near Lake Natron, northern Tanzania, as viewed looking northwards from the slopes of Oldoinyo Lengai is made up of multiple, near-horizontal layers of lava and tephra. The braided drainage system in the valley is indicative of a recently formed land surface (by R.N. Scoon)

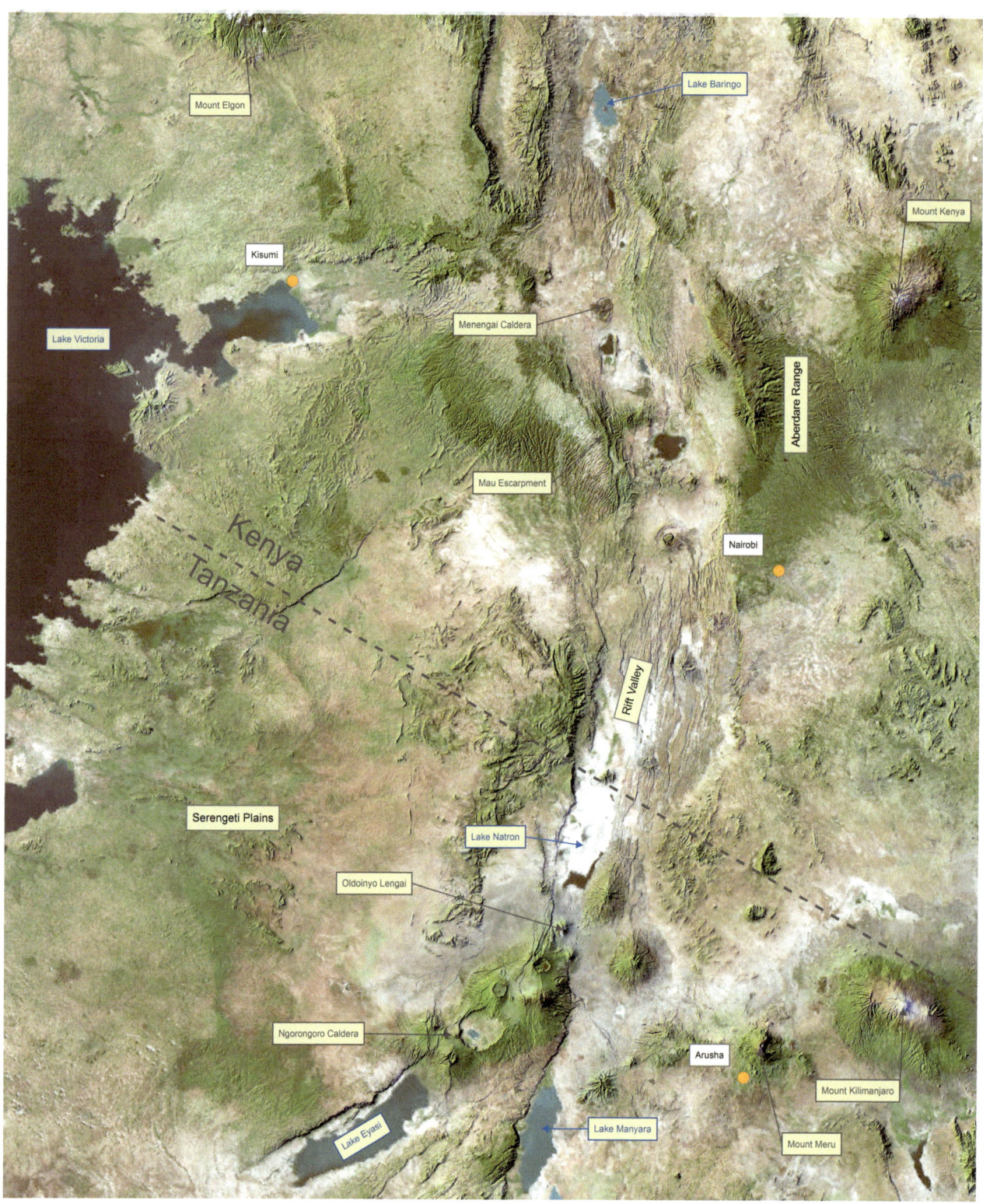

A satellite image of central/southern Kenya and northern Tanzania reveals the great diversity of ecosystems in the region. The two principal geological phenomena are the occurrence of ancient regional plateaus, e.g. the Serengeti and Tsavo Plains and the Gregory Rift. The latter constitutes a narrow linear rift valley bounded by escarpments within Kenya, e.g. the Mau Escarpment, but diverges into several branches in northern Tanzania. A characteristic feature of the rift valley is the chain of ribbon-shaped, alkaline lakes, e.g. Lakes Natron, Manyara and Eyasi. Lake Baringo with its rounded profile and freshwater is unusual. Many of the rift platforms juxtaposed to the rift valley are capped by great thicknesses of plateau-style volcanic outpourings, e.g. those forming the Aberdare Range. Rift platforms also include giant volcanic edifices with successive (annular) botanical zones that include dense montane forests on the lowermost slopes, e.g. Mounts Elgon, Kenya and Kilimanjaro. The explosive style of volcanism that typifies continental rifting has resulted in some cones being capped by giant calderas, e.g. Ngorongoro and Menengai. Active volcanic cones occur both in the rift valley, e.g. Oldoinyo Lengai, and on rift platforms, e.g. Mount Meru.Source: Cloud-free composite image based on Terra-MODIS satellite images of NASA courtesy of Philip Eales (Planetary Visions), with digital terrane shading created using ArcGIS® software by Esri by Oliver Burdekin (birdGIS).

National Parks, Reserves and Conservation Areas

Central and Southern Kenya:

Aberdare; Amboseli; Chyulu Hills; Hell's Gate; Lake Baringo; Lake Bogoria; Lake Naivasha; Lake Nakuru; Mount Longonot; Mount Elgon*; Mount Kenya; Tsavo West

Northern Tanzania:

Arusha (Mount Meru); Kilimanjaro; Lake Eyasi; Lake Manyara; Lake Natron and the Oldoinyo Lengai Volcano; Ngorongoro; Oldupai Gorge and Laetoli; Serengeti

*Includes a park in Uganda.

Preface

The national parks, reserves and conservation areas of central/southern Kenya and northern Tanzania host some of the largest and most diverse concentrations of wildlife remaining on Earth. They also have remarkable potential for **geotourism** and creation of **geoparks**.[1] Many of the most spectacular landforms are associated with the dominant geological feature of the region, the East African Rift System (EARS). Kenya and northern Tanzania are transected by the Gregory Rift, the eastern branch of a system that extends in its entirety for over 6,000 km. The most prominent expression of the rift is a narrow, down-faulted linear valley bordered by spectacular escarpments (Frontispiece 1). The EARS also constitutes one of the largest volcanic provinces on Earth. Sections of the rift valley resemble lunar landscapes in areas of particularly intensive volcanism. An unusual feature of the Gregory Rift is the occurrence of giant volcanic edifices on the rift platforms (Cover). A satellite image (Frontispiece 2) reveals many of these features including the chain of ribbon-shaped lakes which characterise the rift valley.

The rift platforms which constrain the Gregory Rift Valley have been uplifted to elevations sufficiently high as to report a temperate climate with montane forests, despite the equatorial setting. Montane forests also occur on the higher volcanic cones, several of which are among the largest free-standing mountains on Earth. The two highest peaks in the region, Kilimanjaro (a dormant volcano) and Mount Kenya (an extinct volcano), are capped by the relicts of fast-melting icefields and glaciers. The region includes two 'Natural Wonders of the World', the Serengeti Plains, famed for the biannual migration of almost two million grazers, and the Ngorongoro Caldera, a self-contained reserve created by a catastrophic volcanic eruption. The palaeoanthropological sites of Oldupai Gorge and Laetoli, where some of the most important discoveries of hominin fossils and footprints have been made, are located between Ngorongoro and the Serengeti.

The region is underlain in some areas by crystalline metamorphic rocks of the basement complexes. The Serengeti Plains, for example, reveals exposures of the Lake Victoria Terrane which contains some of the oldest rocks on Earth (oldest ages reported in this area are 2.810 Ga). The Tsavo Plains include exposures of the Mozambique Belt, an extensive group of rocks with an average age of 800–500 Ma. A characteristic feature of the East African landscape is the occurrence of regional plateaus; they formed due to repeated episodes of uplift and erosion associated with the breakup of the ancient supercontinent of Gondwana (commencing at approximately 180 Ma). It is these cycles of uplift and erosion which enabled the basement complexes to be exposed on surface. The EARS is a relatively recent event which commenced at approximately 30 Ma. Rifting is stretching and thinning the African Plate. The main periods of faulting in the Gregory Rift are even younger (3.7–1.4 Ma). Continental rifting is invariably associated with the alkaline style of magmas, and volcanism of the EARS is further characterised by highly explosive Plinian-style eruptions. Numerous catastrophic events are preserved

---

[1]**Geopark:** A Geopark is a unified area that advances the protection and use of geological heritage in a sustainable way, and promotes the economic well-being of the people who live there. There are Global Geoparks and National Geoparks (Wikipedia).

in the geological record; these include giant calderas and sector collapses which have generated among the largest debris avalanches recorded. The high sodium content of these magmas, together with restricted catchments and high rates of evaporation, resulted in the extreme alkalinity of many lakes in the Gregory Rift Valley.

The interrelationship between wildlife and geology is emphasised. The relatively youthful volcanic uplands, for example, provide a range of ecosystems and supply groundwater to the much older regional plateaus that would otherwise be far too arid to support large concentrations of wildlife. The intensity of volcanism is not compatible with the generalised view of endless plains and vast herds that migrate over ancient, unchanged landscapes. The geological record provides evidence of persistent changes. Recent eruptions of the active Oldoinyo Lengai Volcano have resulted in the spread of ashes onto the Serengeti Plains which support the nutrient-rich short grasses on which the migrating grazers congregate annually. The episodic nature of the rifting and volcanism produced smaller and smaller geological terranes. This may have resulted in the remarkable diversity of species for which East Africa is so renowned. Anomalously rapid speciation may have been triggered as a reaction to these new terranes as well as to the unusually high concentrations of radioactive minerals. Initiation of new species by Darwinian evolution during the Pliocene and Pleistocene epochs (5.3 Ma–11,500 BP), including hominins, occurred in remarkably short time intervals during extreme climatic cycles (e.g. the Ice Ages) and intense volcanism. The Holocene epoch represents a relatively quiescent period in which *Homo sapiens* thrived during a time of modest climatic cycles and less intense volcanism.

Grahamstown, South Africa   Roger N. Scoon

Acknowledgements

This contribution would not have been possible without the detailed maps produced by the Geological Surveys of Kenya and Tanzania, together with mapping and research reports of the British Geological Survey in East Africa. The research by Louis and Mary Leakey and Richard Hay at Oldupai Gorge and Laetoli, together with the detailed studies of B. H. Baker on the Gregory Rift, Kenya, and by J. B. Dawson on the carbonatite volcanism at Oldoinyo Lengai, northern Tanzania, stands out from the extensive data base. The contribution of Lyn Whitfield who drafted the maps and diagrams was particularly important. Satellite and digital terrane images were provided and processed by Philip Eales (Planetary Vision) and Oliver Burdekin (birdGIS). The 1995 field trip of the Mineralogical Society of South Africa to northern Tanzania planted the seed for the book and this continued with a similar trip led by the author as part of the 35th International Geological Congress, Cape Town (2016). The support of Morris and Richard Viljoen during these excursions, as well as discussions over the ensuing years is greatly appreciated. Reviews by Professor Carl Anhaeusser and M. J. Wilson greatly improved an earlier draft. The support and encouragement of Petra Van Steenbergen, Executive Editor at Springer, was also invaluable. Photographs by the author reflect contrasts between visits during 'dry seasons' (June 1995; September 2016) and 'wet seasons' (January 1980; December 1998; February 2004; January 2012; January 2015). Finally, I thank my wife, Amelia, for accompanying me on many of the field excursions and mountain hikes. We both gratefully acknowledge the private guides and rangers who provided safe guidance and stimulating discussions during our visits to national parks and reserves, particularly whilst trekking on Kilimanjaro (1980), Oldoinyo Lengai (1995), Mount Meru (2004), Mount Elgon (2012) and Mounts Longonot and Kenya (2015).

Contents

Abbreviations

BP   Years before present
DAD   Debris Avalanche Deposit
EARS   East African Rift System
Ga   Billions of Years
Ma   Millions of years
NCA   Ngorongoro Conservation Area
RAMSAR   Conservation on Wetlands signed in RAMSAR, Iran in 1971

List of Figures

List of Plates

List of Boxes

List of Tables

Part I
Overview

Abstract

The national parks, reserves and conservation areas of central/southern Kenya and northern Tanzania reveal a wide range of geological terranes with spectacular landforms. The regional plateaus that are so characteristic of East Africa, e.g. the Serengeti and Tsavo Plains, are underlain by some of the oldest rocks on Earth. The most significant landforms, however, are associated with the East African Rift System (EARS), specifically the eastern branch or Gregory Rift, a relatively recent geological phenomenon that includes intensive volcanism. Rifting and volcanism have dissected the regional plateaus into possibly the most iconic scenery on Earth. The Gregory Rift Valley is a narrow, linear, down-faulted feature, bounded by huge escarpments. A chain of mostly ribbon-shaped, shallow, alkaline lakes occurs in arid, desolate areas of the valley. The rift platforms, however, have been uplifted to elevations sufficiently high as to create a temperate climate with extensive montane forests, despite the equatorial setting. A unique feature is the occurrence of giant volcanic edifices in both the rift valley and on the rift platforms. Some of the volcanoes constitute the largest free-standing mountains on Earth. They are girdled by successive botanical zones of montane forests, heath and moorlands, with icefields and glaciers capping the two highest peaks of Kilimanjaro and Mount Kenya. The Ngorongoro Caldera and the Serengeti Plains are gazetted as natural wonders of the world, the former for its self-contained ecosystem and the latter for the famous biannual migration of several million large grazers. The diverse geology of East Africa is associated with some of the greatest concentrations of wildlife on Earth. The great diversity of species and the rapid speciation that characterises the region are a reaction to the intensity of the rifting and volcanism. The interrelationship between wildlife and geology is a remarkable feature and many of the parks and reserves could be reclassified as geoparks. The region can support extensive geotourism, particularly as there are active volcanic cones, e.g. the Oldoinyo Lengai Volcano and near-lunar volcanic landscapes.

Keywords

Alkaline lakes • Geotourism • Gregory Rift
Ngorongoro • Serengeti • Speciation • Volcanism

Photographs not otherwise referenced are by the author.

© Springer International Publishing AG, part of Springer Nature 2018
R. N. Scoon, *Geology of National Parks of Central/Southern Kenya and Northern Tanzania*, https://doi.org/10.1007/978-3-319-73785-0_1

Plate 1.1 Migration of wildebeest and zebra in the vicinity of Lake Ndutu on the Serengeti Plains

1.1 Introduction

Many of the national parks, reserves and conservation areas of central/southern Kenya and northern Tanzania are world-famous sanctuaries that contain some of the largest and most diverse concentrations of large mammals remaining on Earth (Fig. 1.1). Many of these parks and reserves could be consolidated into larger **geoparks** as the unusual (and in many cases unique) geology has resulted in a remarkable diversity of ecosystems and landscapes. The region is dominated by three discrete geological features and processes, the crystalline basement complexes that include some of the oldest rocks on Earth, the uplifted regional plateaus and the comparatively recent (and still active) EARS. Rifting has dissected the ancient, uplifted basement complexes. The most well-known manifestation of the EARS is the rift valleys, specifically the Gregory Rift in Kenya and northern Tanzania, linear features bounded by regional escarpments (Frontispiece 1). Rifting is accompanied by intensive volcanism and there are a number of active volcanoes in the region, including the giant Mount Meru (Cover). The volcanism at Mount Kenya is extinct, as is that of the Ngorongoro region, but the highest peak of Kilimanjaro (Kibo) is a dormant volcano.

1.2 Wildlife and Geology

The interrelationship between wildlife and geology within central/southern Kenya and northern Tanzania is pronounced. The migration of several million wildebeest, zebra and other grazers on the Serengeti Plains (Plate 1.1), one of the natural wonders of the world, is in part driven by nutritious grasses that selectively grow on ashes derived from the active Oldoinyo Lengai Volcano. Predators in the Serengeti typically spend hot days on koppies, or small hills composed of ancient crystalline rocks where they catch cooling breezes (Plate 1.2a, b). Many of the parks located in the Gregory Rift occur in areas of remarkable landscapes with extensive volcanic terranes. The steep-sided Ngorongoro Caldera, also gazetted as a natural wonder of the world, is a self-contained ecosystem that was created by a catastrophic volcanic eruption several million years ago. The occurrence of ice-capped volcanoes, e.g. Mount Kenya and Kilimanjaro so near the Equator, has long been a source of amazement and the former contains, together with the Aberdare Mountains, unique melanistic species of a number of large mammals. Many parks within the region are able only to sustain the large quantities of wildlife due to groundwater fed from aquifers located in juxtaposed volcanic uplands.

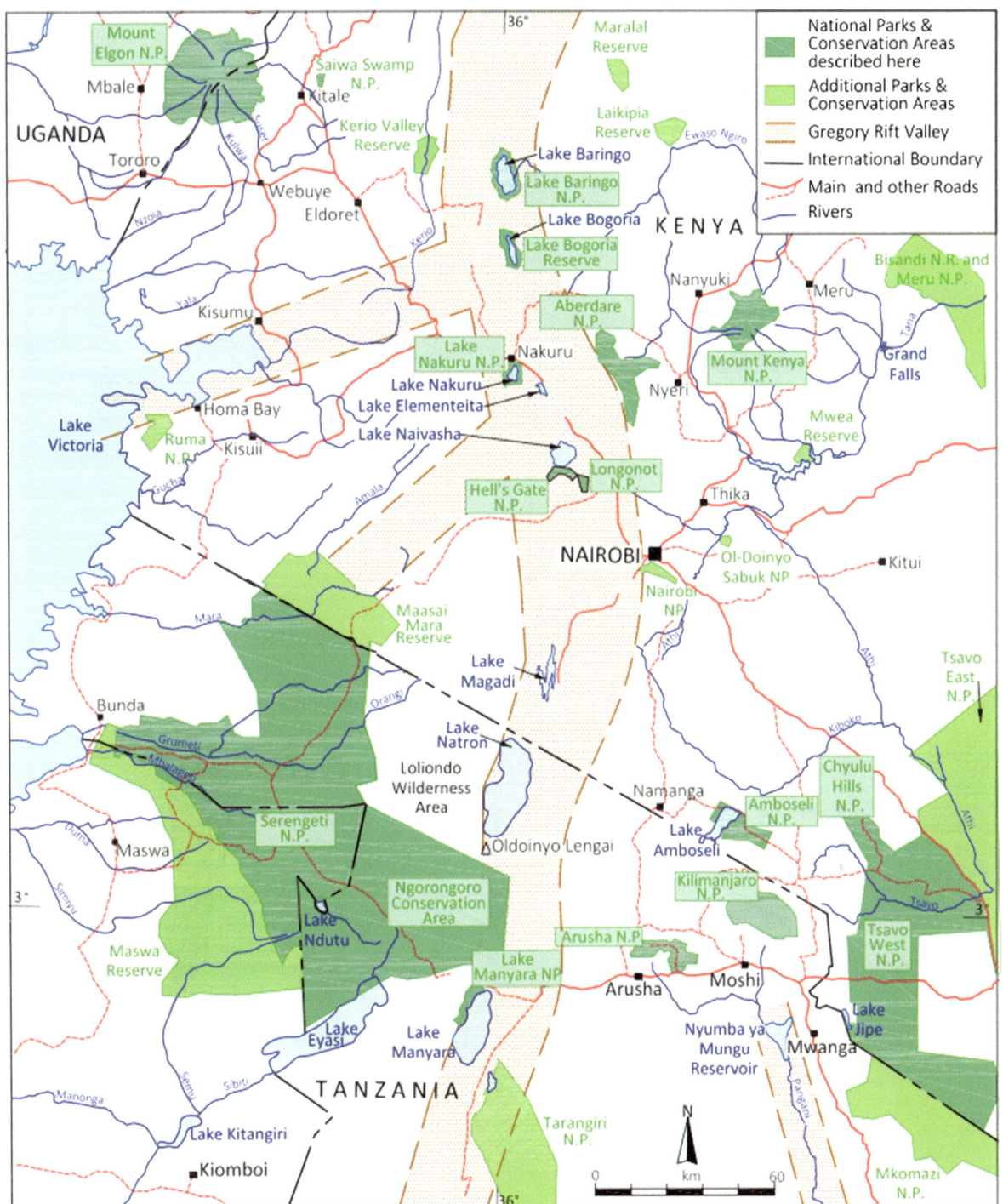

Fig. 1.1 Map of the national parks, reserves and conservation areas of central/southern Kenya and northern Tanzania. Mount Elgon includes a national park in both Kenya and Uganda

1.3 The Gregory Rift

The EARS is comprised of three branches, the Ethiopian, Albertine and Gregory, as depicted on a satellite image (Fig. 1.2). The Ethiopian Rift is the most seismically active of the three rifts and includes the Afar Triangle, one of the most intense areas of volcanism known on Earth. The Albertine (or western) Rift is the most extensive and contains many of the largest lakes in Africa. The Gregory Rift is restricted to Kenya and northern Tanzania. The Gregory Rift and its associated volcanism have influenced the landscapes

Fig. 1.2 A satellite image of East Africa reveals the extent of the three branches of the East Africa Rift System. The Gregory Rift is characterised by a chain of small ribbon-shaped, typically shallow and alkaline lakes, whereas the Albertine Rift includes some of the largest freshwater lakes on Earth. Source: cloud-free composite image based on Terra–MODIS satellite images of NASA, courtesy of Philip Eales, Planetary Visions, with further processing by Oliver Burdekin, birdGIS

of all of the parks and reserves described here. The first geologist to observe and report on the rift valley in Kenya was Joseph Thomson (Thomson 1880), an early pioneer and settler commemorated by, for example, Thomson's Falls near Nyahururu and Thomson's gazelle (*Eudorcas thomsonii*). This branch of the rift is, however, named after John Walter Gregory, a distinguished geologist and intrepid explorer (Gregory 1894a). During his pioneering travels in Kenya, Gregory reported on many features of the rift valley and surrounding areas, including the recent volcanism of the Chyulu Hills, the active Longonot Volcano at the foot of the Mau Escarpment, the alkaline nature of Lake Bogoria and the volcanic structure and icefields of Mount Kenya (Gregory 1894b).

1.4 Land Surfaces

Two principal land surfaces are recognised within central/southern Kenya and northern Tanzania. The oldest surface is the regional plateau that is such a characteristic feature of large parts of the African continent. The regional plateau is typically underlain by ancient basement complexes and reveals open landscapes typified by grassy and bushy savannahs, e.g. the Serengeti and Tsavo Plains. The regional plateau, excluding the later dissection by the EARS, is generally only disrupted by small inselbergs ('Island Mountains') of particularly resistant rocks. The most spectacular landforms in the region, however, are associated with the Gregory Rift. This narrow, linear valley, which is bounded by regional escarpments, hosts a chain of ribbon-shaped lakes. Many of the platforms adjacent to the rift valley have been uplifted to considerably higher elevations than the regional plateau. This has created localised areas with a temperate climate and montane forests, despite the equatorial setting. The uplifting of rift platforms has had pronounced effects on the drainage patterns of the region; most rivers flow *away* from the Gregory Rift, either eastwards into the Indian Ocean or westwards into Lake Victoria (Fig. 1.1). The restricted catchments of the lake basins in the Gregory Rift, together with relatively high rates of evaporation, have resulted in many of the ribbon lakes being strongly alkaline.

Two broad categories of volcanic terranes are associated with the Gregory Rift. Plateau-style or flood events have smoothed out irregularities in both the rift valley and on the rift platforms, the latter being associated with extensive upland areas such as the Aberdare Range. Discrete volcanic cones occur both within the rift valley, where in some areas they are associated with near-lunar landscapes, as well as on the rift platforms where they include some giant volcanic edifices.

1.5 Ecological Regions

The five broad ecological zones that characterise large parts of central/southern Kenya and northern Tanzania are apparent on a satellite image (Frontispiece 2). Each ecological zone correlates with a discrete geological terrane:

- Areas of low relief (pale grey) are associated with arid sections of the rift valley;
- Areas of moderate relief (light brown) include semi-arid regions and deserts located in rain shadows created by the volcanic uplands;
- Savannah grasslands (light green) are mostly restricted to the regional plateau;
- Forested areas (dark green) occur on rift platforms and on the flanks of the large volcanic cones;
- The giant volcanic cones have poorly vegetated upper slopes (pale grey) and may be capped by icefields.

1.6 Distribution of Species

East Africa has great concentrations of wildlife and an unusual diversity of species, particularly in the vicinity of the rift valleys (Williams et al. 1967; Smith 1988). Examples include the abundance of plains animals, occurrence of several species of great apes (Plate 1.2c, d) and extraordinary number of bird species, including the lesser flamingo (*Phoeniconaias minor*) and greater flamingo (*Phoenicopterus ruber*) (Plate 1.2e, f). The localised occurrence of some species is of considerable interest. Some large grazers, for example, are restricted to specific reserves and do not, as may be commonly thought, occur over the entire region, e.g. the Plains or Burchell's zebra (*Equus quagga* or *Equus burchellii*) despite being widespread throughout Tanzania and central/southern Kenya is absent from most of central and eastern Uganda. In comparison, Grevy's zebra (*Equus grevyi*) is restricted to small parts of northern Kenya and Ethiopia. The occurrence of melanistic species of mammals that are restricted to montane forests in only a handful of parks is also significant.

An intriguing feature of areas associated with the Gregory Rift is the absence of the eastern chimpanzee (*Pan troglodytes schweinfurthii*) and mountain gorilla (*Gorilla beringei beringei*), despite localised montane forests (e.g. slopes of many of the volcanic cones; the Ngorongoro Highlands) offering similar habitats to those in parts of Rwanda, the Democratic Republic of Congo, Uganda and western Tanzania. Whether their absence can be ascribed to climatic changes during the past four or five million years (e.g. Demencal 2004), restricted gene pools or human activity is speculative. The opening up of savannah grasslands in response to climatic changes during the past few million years, when forests that previously covered large parts of East Africa retreated westwards, could be a significant factor (e.g. Nicholson 1996; Scholz et al. 2011). The evolution of these species has, however, coincided with much of the geological activity associated with the EARS. Gorillas are thought to have descended approximately nine million years ago from ancestral apes which first appeared in the Oligocene epoch (Sect. 1.2). An early ancestor of the gorilla may have been *Proconsul africanus*, an extinct species of ape which also gave rise to early hominins. Mountain gorillas in turn split off from the gorilla genus approximately two million years ago, a time when several species of hominins, including our human ancestors (*Homo habilis*), inhabited Olduvai Gorge in northern Tanzania. Moreover, the population of mountain gorilla within the Bwindi Impenetrable Forest (ancient basement complex) in Uganda, which is entirely isolated from the larger populations of the Virunga Mountains (active volcanoes of the EARS), may be a separate subspecies (Stanford et al. 2001).

The EARS is still active (e.g. Sarie et al. 2014) and some of the localised variations noted above may have been a reaction to faulting and volcanism within the last few millions of years. Some species may have evolved in isolated geological terranes. The great diversity of species in East Africa is almost unparalleled on Earth, with the exception of the Indonesian and associated islands, most of which are related to relatively recent volcanism.

1.7 Rapid Speciation

A well-known example of the diversity that characterises East Africa is the recognition of numerous species of cichlid fish in the Great Lakes, e.g. more than two thousand occur just in Lake Victoria (Johnson et al. 1996). The speciation of the cichlid fish may be a reaction to the highly variable geologic and palaeoclimatic cycles of the region (Danley et al. 2012). Lake Victoria also has the distinction of being one of the first localities where the annual cycle of phytoplankton in relationship to changes in stratification was first investigated (Talling 1966). East Africa has also revealed fossil evidence of an unusually diverse range of mammal species that became extinct during the past five million years (Demencal 2004). A similar hypothesis of highly variable geologic and palaeoclimatic cycles may apply to the extinct mammals and localised populations of current populations, some of which were noted above. The rapid speciation may have been triggered by the fallout of radioactive volcanic ash associated with the EARS (Sturmbauer et al. 2011;

Ebisuzaki and Maruyama 2015). This hypothesis is consistent with the composition of the alkaline volcanism that is a fundamental component of all continental rifts, including the EARS (Bailey 1974).

Triggering of speciation by radioactivity, together with reaction to climatic cycles and the loss of extensive woody cover over large areas of East Africa, may also have resulted in the evolution of hominins in the vicinity of the East African Rift valleys (e.g. Cerling et al. 2011). Hominin discoveries made at Oldupai Gorge, e.g. *Paranthropus boisei* and *Homo habilis*, are among the most significant ever made (Leakey 1974). The excellent preservation of fossils at Oldupai is in part due to them being buried in volcanic ash derived from the extinct Ngorongoro volcanism. The 27-m-long trail of fossil footprints ascribed to *Australopithecus afarensis* at Laetoli are preserved in volcanic ash from the same area (Leakey 1981). The *Australopithecus* roamed large parts of East Africa over a period of several millions of years and we can assume migration patterns were enforced by annual rainfall patterns, as they are with the grazers on the Serengeti Plains, but also, and more fundamentally perhaps, by faulting and volcanism.

1.8 Layout of the Book

Part I includes, in addition to these introductory comments, an overview of the regional geology (Chap. 2) and of the EARS (Chap. 3) directed at the non-specialist. The geologist or interested layperson will find more details of the regional geology in Part II, i.e. Basement complexes and regional plateaus (Chap. 4), the Gregory Rift (Chap. 5), and the Late Pleistocene Ice ages and the Holocene epoch (Chap. 6). Part III summarises the geology of the most well-known parks and reserves in the region. The sequence is based on the ages of the dominant geological terrane. The oldest rocks in the region underlie the Serengeti Plains (Chap. 7) and the oldest of the giant volcanic edifices is Mount Elgon (Chap. 8). The Aberdare Range and Mount Kenya are dominated by extinct volcanic terranes (Chap. 9), as is the Ngorongoro Conservation Area (NCA) (Chap. 10). The world-famous palaeoanthropological sites of Oldupai Gorge and Laetoli occur in the NCA (Chap. 11). Kilimanjaro is a giant volcanic edifice, part of which is dormant, capped by icefields which reveal important details of climatic changes during the Holocene epoch (Chap. 12). The Arusha National Park, possibly the least well known and yet one of the most spectacular parks in East Africa, is dominated by Mount Meru, an active volcano (Chap. 13). The arid plains of the Amboseli and Tsavo West National Parks are fed groundwater from the volcanic uplands of the Chyulu Hills National Park and include several recent lava flows (Chap. 14). Many of the lakes in the rift valley occur in desolate terranes (Chap. 15) and the Hell's Gate and Mount Longonot National Parks, also located in the rift valley, include active volcanoes located next to Lake Naivasha (Chap. 16). Oldoinyo Lengai is the most persistently active volcano in the region, located in a spectacular setting near Lake Natron (Chap. 17). A brief glossary of geological terminology is appended.

Plate 1.2 East Africa reveals a diversity of wildlife. Predators such as lion (**a**) and cheetah (**b**) are often found on the flanks of small hills (or koppies) on the Serengeti Plains, northern Tanzania. Primates such as mountain gorilla in Rwanda (**c**) and chimpanzee in Uganda (**d**) are restricted to forests in the vicinity of the Albertine Rift. The lesser flamingo (pink) mixes with the greater flamingo (white) on Lake Elmenteita, Kenya (**e**) with the latter being particularly abundant on Lake Nakuru, Kenya (**f**). Photographs (**e**) and (**f**) courtesy of Peter Prokosch from the album African Diversity

References

Bailey, D. K. (1974). Continental rifting and alkaline magmatism. In H. Sorensen (Ed.), *The alkaline rocks* (pp. 148–159). New York: Wiley.

Cerling, T. E., Wynn, J. G., & Andanje, S. A. (2011). Woody cover and hominin environments in the past 6-million years. *Nature, 476* (7358), 51–56.

Danley, P. D., Husemann, M., Ding, B., DiPietro, L. M., Beverly, E. J. & Peppe, D.J. (2012). The impact of the geologic history and paleoclimate on the diversification of East African cichlids. *International Journal of Evolutionary Biology 2012* (20 p), Article ID 574851.

Demencal, P. B. (2004). African climate change and faunal evolution during the Pliocene-Pleistocene. *Earth and Planetary Science Letters, 220*(1–2), 3–24.

Ebisuzaki, T., & Maruyama, S. (2015). United theory of biological evolution: disaster-forced evolution through Supernova, radioactive ash fall-outs, genome instability, and mass extinctions. *Geoscience Frontiers, 6*, 103–119.

Gregory, J. W. (1894a). Contributions to the physical geography of British East Africa. *Geographical Journal*, 4, 290–315, 408–424, 505–514.

Gregory, J. W. (1894b). Contributions to the geology of British East Africa: glacial geology of Mount Kenya. *Quarterly Journal Geological Society of London, 50*, 515–530.

Johnson, T. C., Scholz, C. A., & Talbot, M. R. (1996). Late Pleistocene desiccation of Lake Victoria and rapid evolution of cichlid fishes. *Science, 273*(5278), 1091–1093.

Leakey, L. S. B. (1974). *By the evidence: memoirs 1932–1951* (276 p). New York: Harcourt, Brace, Jovanovich.

Leakey, M. D. (1981). Discoveries at Laetoli in Northern Tanzania. *Proceedings of the Geologists' Association, 92*(2), 81–86.

Nicholson, S. E. (1996). A review of climate dynamics and climate variability in Eastern Africa. In T. C. Johnson & E. O. Odada (Eds.), *The limnology, climatology and paleoclimatology of the East Africa Lakes* (pp. 25–56). Amsterdam: Gordon and Breach.

Sarie, E., Calais, E., Stamps, D. S., Delvaux, D., & Hartnady, C. (2014). Present day kinematics of the East African Rift system. *Journal of Geophysical Research, 119*, 3584–3600.

Scholz, C. A., Cohen, A. S., Johnson, T. C., King, J., Talbot, M. R., & Brown, E. T. (2011). Scientific drilling in the Great Rift Valley: the 2005 lake Malawi drilling project—an overview of the past 145,000 years of climate variability in Southern Hemisphere East Africa. *Paleography, Paleoclimatology, Paleoecology, 3030*(1–4), 3–19.

Smith, A. (1988). *The Great Rift: Africa's changing valley* (224 p). London: BBC Books.

Stanford, C. (2001). The subspecies concept in Primatology: the case of mountain Gorillas. *Primates, 42*(4), 309–318.

Sturmbauer, C., Husemann, M. & Danley, P. D. (2011). Explosive speciation and adaptive radiation of East African cichlid fishes. In: F. E. Zachos & J. C. Habel (Eds.), *Biodiversity Hotspots-distribution and protection of conservation priority areas* (pp. 333–362). Amsterdam: Springer.

Talling, J. F. (1966). The annual cycle of stratification and phyto-plankton growth in Lake Victoria (East Africa). *International Review of Hydrobiology, 51*(4), 545–621.

Thomson, J. (1880). Notes on the geology of East-Central Africa. *Nature, 28*, 102–104.

Williams, J. G., Arlott, N. & Fennessy, R. (1967). *Collins field guide to National Parks of East Africa* (336 p). Hong Kong: Harper Collins.

Abstract

The geological framework of East Africa is comprised of three terranes: basement complexes, regional plateaus and the East African Rift System (EARS). The oldest of the basement complexes is the Archaean-age Lake Victoria Terrane, part of the extensive Central African craton which dominates the area between the Albertine and Gregory Rifts. The Lake Victoria Terrane includes greenstones, altered sedimentary and basaltic rocks, as well as large plutons of granite-gneiss. The Neoproterozoic-age Mozambique Belt is restricted to areas east of the craton and includes both metasedimentary rocks and granitic plutons. Preservation and exposure of these ancient, crystalline rocks is ascribed to cycles of uplift and erosion associated with the break up of the supercontinent of Gondwana. This process started during the Jurassic at approximately 180 Ma. The most pronounced phase of erosion produced the 70-Ma-old Cretaceous-age African Surface. The EARS commenced in Ethiopia in the Late Oligocene (at approximately 30 Ma) and propagated southwards reaching Kenya in the Miocene and northern Tanzania in the Pliocene. The principal manifestations of the EARS are well-defined, narrow, linear valleys enclosed by elevated rift platforms, together with intensive volcanism. Sedimentary basins are a localised feature of the EARS but are important as they may contain fossils of both hominins and extinct mammals. Hominins evolved during the Pliocene and Early Pleistocene of East Africa in epochs of intensive tectonism and volcanism. The Late Pleistocene of East Africa confirms that the Ice Ages were global phenomena. *Homo sapiens* have thrived in the Holocene, i.e. since 11,500 BP in an epoch of comparatively modest climatic cycles and less intense volcanism.

Keywords

African surface • Basement complexes • EARS Hominins • Regional plateaus • Volcanism

Photographs not otherwise referenced are by the author.

© Springer International Publishing AG, part of Springer Nature 2018
R. N. Scoon, *Geology of National Parks of Central/Southern Kenya and Northern Tanzania*, https://doi.org/10.1007/978-3-319-73785-0_2

Plate 2.1 The spectacular volcanic edifices of Kibo (foreground) and Mawenzi are the two highest peaks within the Kilimanjaro massif. Kibo is capped by a collapsed caldera that includes the dormant Reusch Crater. Photograph from public domain website http://2.bp.blogspot.com/summit_kilimanjaro.jpg

2.1 Introduction

The dominant geological and geomorphological features of East Africa are the ancient crystalline basement complexes, the regional plateaus and the East African Rift System (EARS), the latter including both the well-known rift valleys (bounded by regional escarpments) and the uplifted rift platforms. It is possibly less well known, however, that the EARS is also associated with intensive volcanism. The East African volcanic province includes active systems. A particular feature is the occurrence of giant volcanic cones, none more spectacular than Kilimanjaro (Plate 2.1). A simplified geological map shows the distribution of the more important geological systems in East Africa (Fig. 2.1). The principal rock systems are represented in a stratigraphic column (Fig. 2.2). Also shown here are the more important geological processes which affected earlier-formed rocks, notably the uplift and erosion that occurred in association with the break up of the ancient supercontinent of Gondwana, as well as the EARS.

Ages are expressed in either billions of years (Ga) or millions of years (Ma) for older events and years before present (BP) for those younger than 100,000 years. We are currently experiencing the Holocene epoch which commenced at 11,700 BP when a rapid explosion in population numbers of *Homo sapiens* occurred as the Late Pleistocene Ice Ages ended.

The three distinctive groups of rocks and or geological features that characterise East Africa are as follows:

- Basement complexes (Archaean–Proterozoic Eras);
- Regional plateaus (Jurassic–Neogene periods);
- The rifting and volcanism of the EARS (Oligocene–Holocene epochs).

2.2 Basement Complexes

The basement complexes of East Africa encompass almost two billion years of the geological record, prior to the appearance of hard-shelled animals that mark the beginning

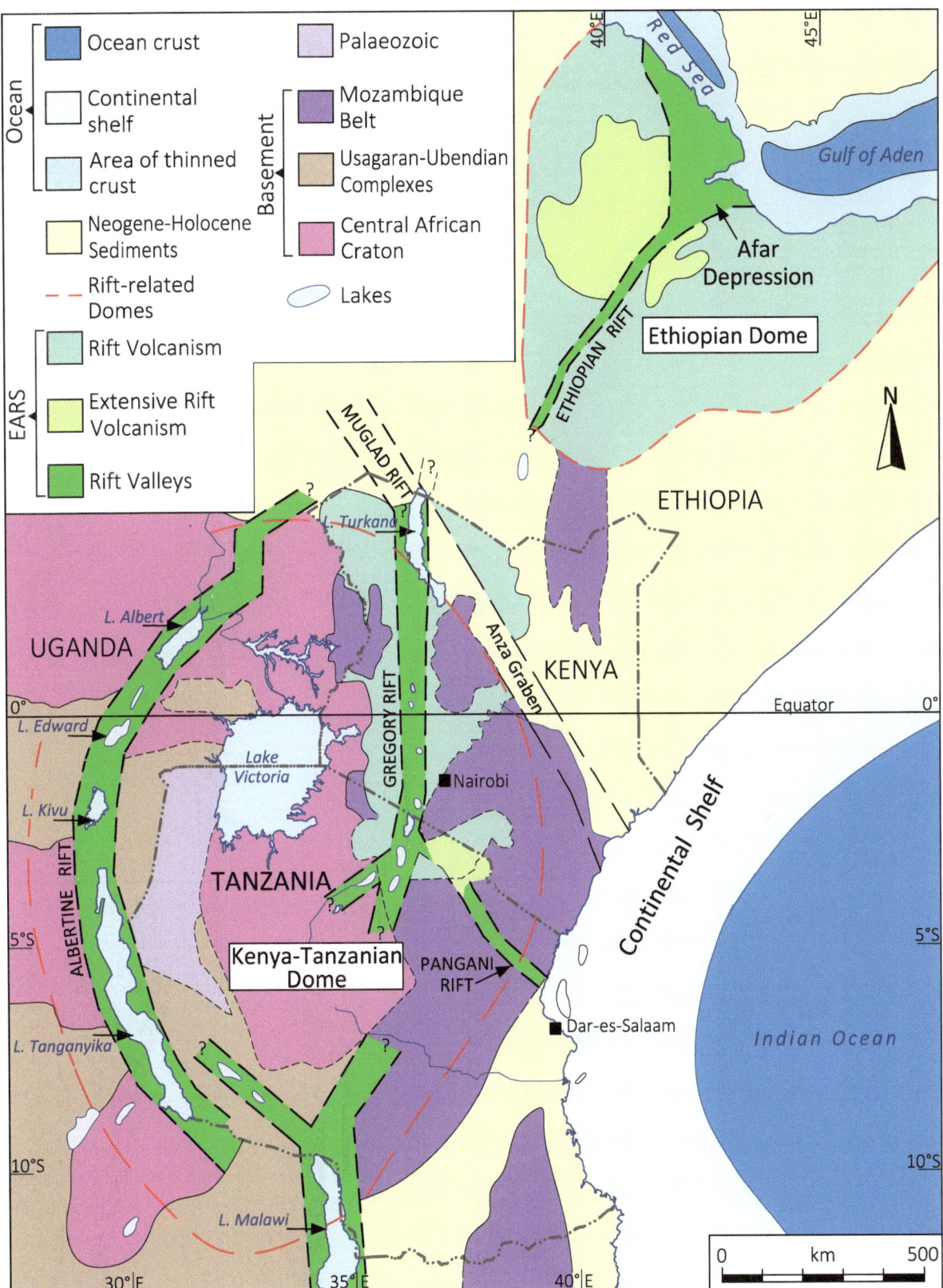

Fig. 2.1 Simplified geological map of East Africa compiled from various sources including the 1:1,000,000 scale Geological Map of Kenya and the 1:2,000,000 scale Geological Map of Tanzania. The three branches of the EARS, Ethiopian, Gregory and Albertine, are truncated by older, Cretaceous-age rifts. Rift valleys contain extensive sequences of Neogene-age volcanic rocks and localised sedimentary basins. Rift platforms to the Ethiopian and Gregory Rifts are dominated by Neogene-age volcanics

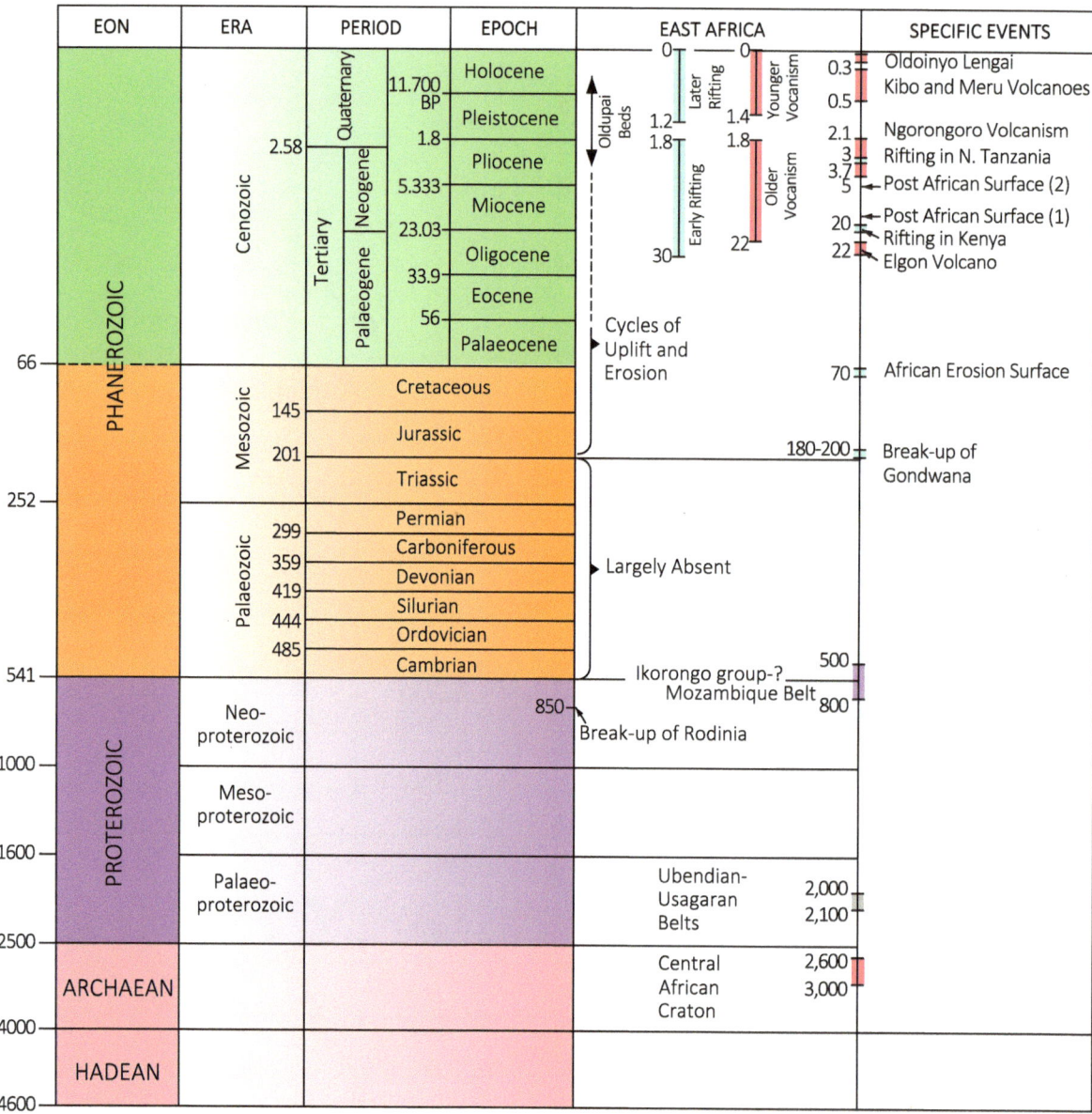

Fig. 2.2 Simplified stratigraphic column for East Africa. Ages (Ma) incorporate latest recommendations of the International Union of Geological Sciences with the exception of the Pliocene–Pleistocene boundary which is retained at the traditional age of 1.8 Ma (not 2.58 Ma) as this would impact severely on published literature pertaining to many of the sites in northern Tanzania, including Oldupai Gorge

of the Phanerozoic. The oldest component is the Central African craton. This stable massif is comprised of Archaean-age cratonic nuclei (greenstones and granite-gneisses) welded together by the Usagaran and Ubendian complexes. The latter are extensive Palaeoproterozoic belts dominated by quartzites, schists and granites that have experienced repeated cycles of metamorphism and deformation. There are no Mesoproterozoic rocks preserved in the region. The youngest of the basement complexes in East Africa is the Mozambique Belt, a north–south trending Neoproterozoic mobile belt comprised of quartzites, schists and granitic plutons.

2.3 Regional Plateaus

The preservation and exposure of the basement complexes in East Africa is ascribed to repeated phases of continental-scale uplift and erosion which accompanied the breakup of the ancient supercontinent of Gondwana. Some regionally important structural lineaments (typically aligned north–northeast) within the region are associated with this process. The African Plate separated from Gondwana with formation of the Atlantic and Indian Oceans. The other components of the supercontinent, including South America, India and Australia, drifted away from Africa as rifting persisted. The breakup commenced in the Jurassic period at approximately 180 Ma and persisted episodically into the Paleogene. The paucity of rocks from the Palaeozoic and Mesozoic Eras in East Africa may be because they were removed by erosion accompanying the uplift. Erosion was particularly intense during the humid climate of the Cretaceous and Palaeogene. The 70-Ma-old African Surface (Late Cretaceous) is the most widely developed erosion surface and is associated with many of the regional plateaus in East Africa.

2.4 East African Rift System

The most iconic landscapes in East Africa are associated with the EARS. The EARS includes three discrete geological processes: faulting, volcanism and formation of small sedimentary basins. Faulting and volcanism are still active in many parts of the region and sedimentary basins are persistently being reshaped by down-warping and erosion of the adjacent scarps. The Kenyan-Tanzanian and Ethiopian Domes, which in addition to the rift valleys are the most pronounced consequence of the EARS, are regional features. The Gregory Rift reveals extensive volcanic terranes. Volcanism includes both plateau-style outpourings—lavas that flood the older land surfaces, smoothing out the palaeo topography—as well as giant, discrete cones, such as Mount Kenya and Kilimanjaro. The Marsabit region, Kenya displays an unusual landscape created by clusters of relatively small, individual cones (Plate 2.2a). Volcanic craters, many containing small lakes, are a particular feature of the Queen Elisabeth National Park and surrounding areas in Uganda (Plate 2.2b).

The regional escarpments that define the rift valleys are associated with the main phase of north–south faulting. Escarpments demarcate the edge of rift platforms (or shoulders) which have, in some localities been built up to tremendous heights by plateau-style volcanic outpourings. Some rift platforms are located more than a thousand metres higher than regional plateaus. Moreover, giant volcanic edifices occur not only in the rift valleys but also on the rift platforms.

Sedimentary basins associated with rifting are localised features that are restricted to either the rift valleys or small warps on rift platforms. The sedimentary basins are mostly infilled by sandstones, clays and volcaniclastics derived from erosion of the adjacent scarps and plateaus. Many of the ribbon lakes of the Gregory Rift are located in small basins at the base of regional escarpments, e.g. Lake Elmenteita, Kenya (Plate 2.3a).

2.5 Hominins

Sedimentary basins associated with the EARS provide favourable localities for the preservation of hominin fossils. Hominins evolved during the Pliocene and Early Pleistocene of eastern and southern Africa when the climate was comparatively hot and humid. They were displaced by *Homo sapiens* during the Late Pleistocene at approximately 0.2 Ma.

2.6 Late Pleistocene Ice Ages and the Holocene Epoch

Extensive icefields, including slope glaciers developed during the Late Pleistocene Ice Ages on all of the significant peaks in East Africa, e.g. Kilimanjaro, Mount Elgon, Mount Kenya, Mount Meru and the Ruwenzori (Uganda and the Democratic Republic of Congo). In comparison, the Holocene has been a relatively quiescent epoch with modest climatic cycles and volcanism, albeit recent warming is causing the ice sheets and glaciers on Kilimanjaro to recede at an alarming rate (Plate 2.3b). Holocene-age sediments and volcanic ash occur in some of the sedimentary basins. The surficial deposits that dominate the coastal plains of East Africa, i.e. at the foot of the escarpment that defines the eastern extremity of the regional plateaus, are mostly of Holocene age.

Plate 2.2 a Clusters of volcanic cones in the Marsabit Region of northern Kenya are located on the eastern platform of the Gregory Rift; **b** One of a group of volcanic craters with sodas lakes on the Kazinga Plains, Queen Elisabeth National Park, Uganda

Plate 2.3 a Lake Elmenteita, Kenya is located in the Gregory Rift beneath the Mbaruk Escarpment, Kenya. Photograph by Peter Prokosch from the album African Diversity; **b** View of the Furtwängler Glacier, the last ice sheet entirely located on the summit plateau of Kibo (Kilimanjaro), looking north towards the Northern Ice field over the Alpine desert. *Source* Public domain website https://upload.wikimedia. org/wikipedia/commons/a/a1/Glacier_at_summit_of_Mt_Kilimanjaro_ 001.JPG (2012)

2.7 Discussion

Without the Gondwana-related uplift, the interior of East Africa would have been dominated by low-lying tropical forest or desert, environments less suitable for both human evolution and habitation. Moreover, the regional escarpment that separates the interior plateau from the narrow coastal plains has resulted in an absence of navigable rivers that could have facilitated exploration. Exploitation of interior plateaus at an earlier stage of history would probably have decimated the large concentrations of wildlife and severely affected landforms that are far more susceptible to human interference than those on most other continents. East Africa is one of the few areas remaining on Earth where recent evolution patterns (hominins and wildlife) can be investigated. Much of the region is very sensitive to climatic changes and human encroachment.

The East African Rift System

Abstract

The East African Rift System (EARS) is driven by extension and thinning of the African Plate. Rifting commenced during the Oligocene (at approximately 30 Ma) and has persisted intermittently through the Miocene, Pliocene and Pleistocene into the Holocene. Rifts are invariably associated with intense volcanism and East Africa includes one of Earth's largest volcanic provinces. The occurrence of active cones is consistent with a long-lived mantle plume that may have rejuvenated the older Gondwana-age plume. Major seismic activity in Africa is uncommon but moderate-sized earthquakes can be triggered by active faults. Three discrete branches of the EARS are recognised: Ethiopian, Albertine and Gregory. The Ethiopian Rift is part of an active triple junction that includes the Red Sea and Gulf of Aden. Rifting propagated southwards from Ethiopia, synchronously forming the western or Albertine Rift, the most extensive branch, and the smaller, eastern or Gregory Rift. Rifting involves three distinct stages; pre-rift, half-graben and full graben. The pre-rift stage is manifested by regional doming and includes an early phase of intense volcanism. The half-graben stage includes joining of isolated warps into linear features. The development of a rift valley, or full graben, i.e. a down-faulted block enclosed by uplifted rift platforms on either side, is one of the most intriguing of geological features. Rifting has had a pronounced effect on the geomorphology of East Africa. The drainage patterns, including major rivers such as the Nile, have been severely affected. A notable difference between the Albertine and Gregory Rifts is that whereas the former contains some of the world's largest and deepest freshwater lakes, the latter is far more arid and is characterised by a chain of mostly small and shallow, alkaline lakes.

Keywords

Faults • Grabens • Lakes • Magma plume
Rifting • Sedimentary basins • Volcanism

Photographs not otherwise referenced are by the author.

© Springer International Publishing AG, part of Springer Nature 2018
R. N. Scoon, *Geology of National Parks of Central/Southern Kenya and Northern Tanzania*, https://doi.org/10.1007/978-3-319-73785-0_3

Plate 3.1 Linear escarpments associated with the boundary faults of the rift valley in southern Kenya can be traced for tens of kilometres. Photograph courtesy of Peter Prokosch from the album African Diversity

3.1 Introduction

Rifting is one of the underlying principals of plate tectonics as it explains the dismantling of continents and development of oceans. The African Plate, for example, was isolated from the ancient supercontinent of Gondwana due to rifting of the Atlantic and Indian Oceans (Box 3.1). Rifting caused the other continental masses which made up Gondwana, Antarctica, South America, India, Madagascar and Australia to 'drift' apart. Most rifts occur within oceanic basins, but the EARS is an example of a continental rift (Merle 2011). The EARS is stretching (thinning) and pulling apart the continental crust associated with a large segment of the African Plate. The EARS is related to an episodic process which commenced after the breakup of Gondwana, probably during the Oligocene epoch (Baker et al. 1972; Morley et al. 1999; Chorowicz 2005). The EARS has persisted into the Holocene epoch and is considered by geologists as an active system (Sarie et al. 2014). One manifestation of the EARS is the development of linear valleys that are bordered by regional escarpments, e.g. as seen in the Gregory Rift of southern Kenya (Plate 3.1).

The EARS is comprised of different segments that in their extremity extend for some 6,500 km from Lebanon in the Middle East, through the Red Sea and Afar region of Ethiopia, prior to petering out in southern Africa. The Ethiopian Rift is one branch of a triple junction that includes the Red Sea and Gulf of Aden, regions of drastically thinned oceanic crust (Fig. 2.1). The most extensive branch of the EARS is the Albertine Rift which cuts through Sudan, Uganda, the Democratic Republic of Congo, Rwanda and Burundi, western and southern Tanzania, Malawi and central Mozambique. In comparison, the Gregory Rift has a relatively restricted occurrence within Kenya and northern Tanzania. The rift in northern Tanzania divides into three branches, of which only the Pangani Rift extends southwards more than a few tens of kilometres. Possible linkages between the three rifts in East Africa have been discussed by Ebinger et al. (2000).

3.2 Historical Descriptions

The description of the Kenyan section of the rift valley by Gregory (1896) as 'a linear valley with parallel and almost vertical sides, which has fallen owing to a series of parallel faults' is still accurate. Gregory continued his hypothesis by suggesting the valley descended in blocks, typically some tens of kilometres in length, an explanation consistent with modern interpretations of an elastic (rather than rigid) continental crust. This hypothesis is also consistent with observations that the regional escarpments in Kenya are typically restricted to 20–30 km sections and are not joined into one continuous feature (as is incorrectly shown on many regional maps).

The terms 'graben' and 'horst' to describe the rift valleys and rift platforms, respectively, were proposed by Austrian geologist Eduard Suess (Suess 1885–1901). Despite never visiting Africa, Suess was the first to suggest that the unusual arrangement of lakes in East Africa is controlled by linear faulting, an observation applicable to both the Albertine and Gregory Rifts. Not all geologists who visited Africa in the nineteenth century were as prescient as Thomson, Gregory and Suess, however, as in 1852 Roderick Murchison (President of the Royal Geographical Society), pronounced that sub-Saharan Africa was a land of great geological antiquity in which nothing had happened for millions of years.

Box 3.1: Rifting and Continental Drift

The hypothesis of plate tectonics is consistent with observations that individual plates can drift, collide and stretch (e.g. Windley 1977). Rifting is one of the key processes in the hypothesis as is the recognition that continental and oceanic crusts are separate phenomena. Rifts develop as a consequence of stretching and thinning of the Crust (Wilson 1973). Continents can rift so extensively as to form new oceans. Examples are the Atlantic and Indian Oceans which formed during the breakup of the supercontinent of Gondwana. The separation of the island of Madagascar from the African Plate occurred as part of this process, around 100 Ma, albeit within a relatively short time frame. The EARS is the largest and most well-preserved continental rift system on Earth and may be considered as an arm of the Red Sea Rift that has either failed or is temporarily quiescent. Geological processes invariably include long periods of quiescence separated by much shorter lived catastrophic bursts of energy, so whether the EARS will eventually divide the African Plate to create a new ocean is not known. Whereas rifting associated with the Gondwana event involved extensions of thousands of kilometres, the Albertine and Gregory Rifts have revealed maximum lateral extensions of 30 and 8 km, respectively. The most active area within the Red Sea experiences extension of approximately 2.5 cm/annum (or 25 km over 1 Ma).

3.3 Geomorphology

The first phase of uplift of the African Plate was initiated during the Jurassic breakup of the supercontinent of Gondwana. Uplift in East Africa was reactivated during the EARS to form the high plateaus and mountains collectively known as the Ethiopian and Kenya-Tanzanian Domes (Baker et al. 1972). This has resulted in formation of some of the highest plateaus in Africa and the distinction between regional plateaus (Gondwana) and rift plateaus (EARS) is significant. The highest of the EARS-related plateaus are ascribed to thick outpourings of flood basalt (Bailey 1974). Moreover, the relative youthfulness of many of the volcanic cones has tempered the effects of erosion in comparison to the older terranes.

The average width of the Gregory Rift in Kenya is approximately 40–65 km (Baker et al. 1972), but areas of the Ethiopian Rift, most notably in the Danakil region, are considerably wider (up to 480 km). Rifts are defined by a series of major faults that can be traced for several tens of kilometres on surface. Some sections of the rifts are asymmetric: they are half-grabens with major faults restricted to one side. The surface expression of the major faults correlates with regional escarpments, e.g. the 1,000-m-high Mau Escarpment in central Kenya (Frontispiece 2).

Rifting has severely impacted on drainage patterns within the heart of the African continent. The Albertine Rift captured the upper parts of the Nile River which flows westwards (rather than northwards) from Lake Victoria into Lake Albert, as described by Beadle (1981). The lower sections of the Nile are constrained within an older Gondwana-related structural lineament. The Congo River was probably also affected by the Albertine Rift and may originally have flowed eastwards (Stankiewicz and De Wit 2006).

The Albertine Rift contains numerous large and deep freshwater lakes. Many are used to designate international boundaries. The presence of deep lakes has the effect of obscuring the remarkable depth of the rift, e.g. Lake Tanganyika, which occurs in a mountainous terrane, has a depth of 1,470 m (Sander and Rosendahl 1989). In comparison, there are few major rivers within the Gregory Rift and lakes are smaller, shallower and are mostly alkaline (Gregory 1896). This is in part due to the semi-arid environment in which evaporation exceeds ingress, but crucially most catchments are relatively small as rivers flow away from the rift as noted previously. Rivers to the east of the rift in central and southern Kenya, such as the Tana and Tsavo, flow into the Indian Ocean (Fig. 1.1). Rivers to the west of the Gregory Rift in central Kenya and northern Tanzania, the latter including the Mara and Grometi (they can be considered as headwaters of the Nile River), drain into Lake Victoria.

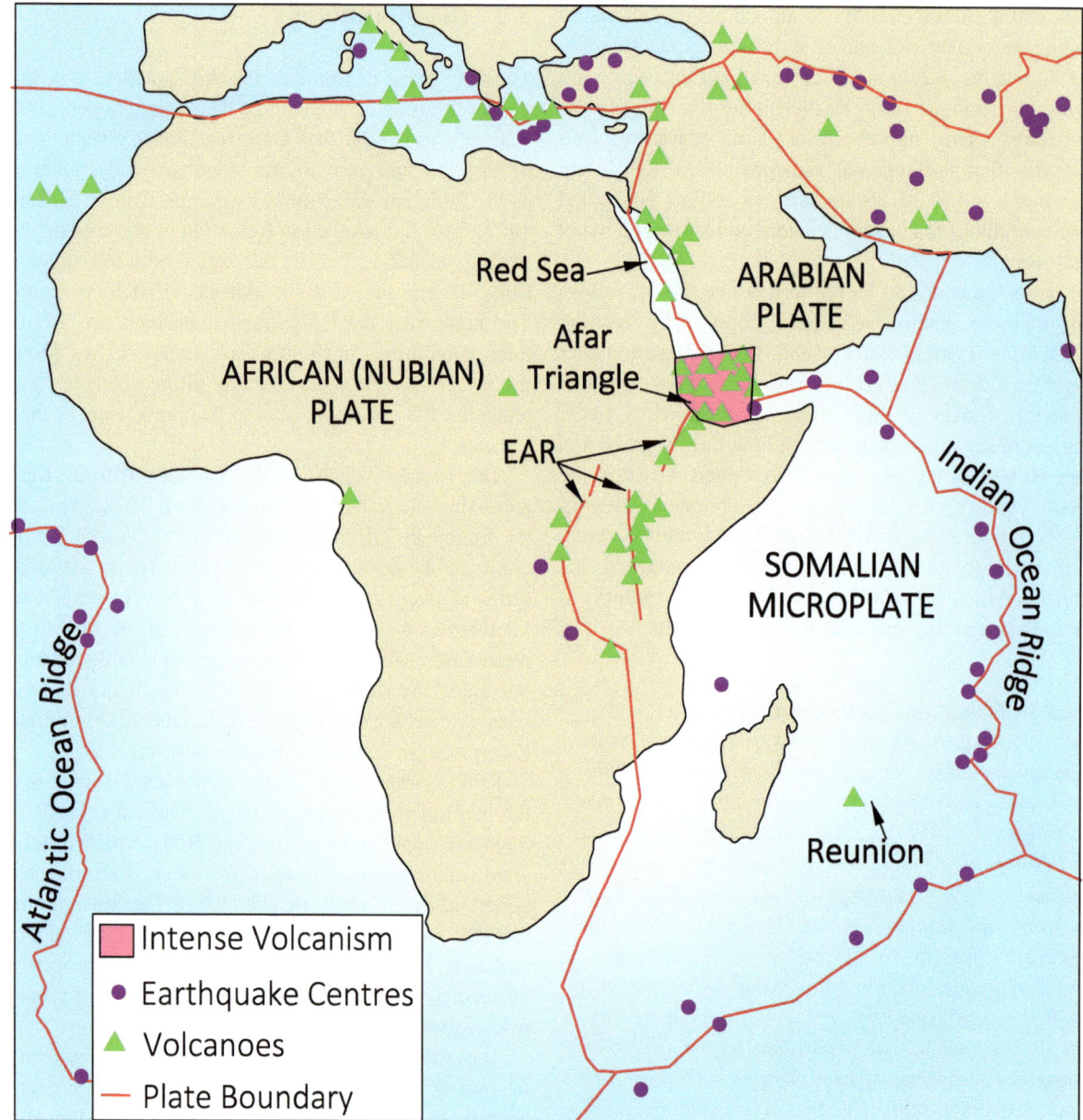

Fig. 3.1 Sketch map of the African Plate showing the boundaries and extent of the East African Rift System. Major seismic events and active volcanoes are aligned with plate boundaries and rifts

3.4   Seismic Activity

The ongoing rifting is causing the African Plate to potentially separate into two plates, a western or Nubian Plate and an eastern or Somalian Microplate (e.g. Ebinger 2005). Most seismic events recognised in and around Africa occur either along the boundary of the existing plate or along active sections of the EARS (Fig. 3.1). The distribution of active volcanoes follows a similar pattern. The valley floor near Murchison Falls, Uganda, where the Nile River plunges into the Albertine Rift (Plate 3.2), has sunk some 4,000 m during the past 14 Ma. There have been several large earthquakes in this area (Maasha 1975). In recent years, a magnitude 7 event in 1966 resulted in 157 fatalities in the Semliki Valley and a magnitude 6 event in 1995, caused a 20-km-long crack to appear in the Fort Portal region (Roberts 2007). Earthquakes in the Gregory Rift are both rarer and of lower intensity.

3.5 Development and Propagation of the EARS

Three stages are generally recognised for development of continental rifts: pre-rift, half-graben and full graben (Merle 2011). The EARS has followed a similar pattern (Baker et al. 1972). Evidence from the pre-rift stage of the EARS demonstrates that regional doming and volcanism preceded the main phases of faulting (Fig. 3.2a). The intermediate half-graben stage triggered asymmetrical faulting and intensive volcanism. The full graben stage is associated with the largest faults and resulted in formation of near-symmetrical valleys (Fig. 3.2b). Pre-rift activity, i.e. regional doming and volcanism, was initiated in the Ethiopian Rift during the Oligocene, at approximately 30 Ma. Rifting propagated southwards to form the Albertine and Gregory Rifts. Pre-rift stages commenced in the Late Oligocene–Early Miocene followed by half-graben (Late Miocene) and full graben stages (Pliocene–Pleistocene).

3.6 Rift Volcanism and Magma Plumes

The Ethiopian and Gregory Rifts are associated with intense volcanism and are known as high-output rifts. The Afar region of Ethiopia includes enormous thicknesses of volcanic rocks that have built up from repeated eruptions over the last 30 Ma. There are a dozen or more active volcanoes in the Gregory Rift, including Oldoinyo Lengai in northern Tanzania. Volcanism is less intense in the Albertine Rift although the Nyiragongo Volcano in the Democratic Republic of Congo is part of a group of active cones within the Virunga range. The relationship between earthquake epicentres and volcanoes associated with the EARS is apparent from Fig. 3.2. Nusbaum et al. (1993) identified 615 volcanoes in the EARS and noted that, whereas the majority of the Holocene (active) volcanoes occur within the rifts, the older (extinct) centres are far more widely scattered.

The intensity of volcanism associated with the Gregory Rift has been widely discussed (e.g. Baker et al. 1972; Dawson 2008) and is one reason why the EARS is thought to be driven by a deep-seated mantle plume (Chorowicz 2005; Ebinger 2005). The rift valleys are infilled by extensive volcanic terranes, with both flood or plateau-style outpourings and discrete cones. The spread of volcanism onto the rift platform is a notable feature of the Ethiopian and Gregory Rifts. This can only be explained by a major thermal event. Volcanism associated with the EARS is so intense and so widely distributed that the existence of two mantle plumes has been postulated (Rogers et al. 2000). Mantle plumes constitute an upwelling of abnormally hot sections of the asthenosphere in which the plume head reacts with the lithosphere triggering partial melting and formation of magma chambers. The longevity of mantle plumes is well known and the EARS may be related to a rejuvenation of a plume associated with the breakup of Gondwana. The continued existence of magma chambers underlying some volcanoes in the Gregory Rift is sufficient for exploitation of geothermal fields for driving steam turbines for production of electricity, such as at the Olkaria Volcanic complex in the Hells Gate National Park, Kenya (Plate 3.3).

3.7 Sedimentary Basins and Lakes

Tectonism associated with continental rifts results in the formation of small basins within rifts, as well as warps on rift platforms (Baker et al. 1972; Dawson 2008). Basins and warps are typically infilled by sediments (mostly sands and clays), volcanic ash deposits and volcaniclastics (reworked lavas and ashes). Some of the sedimentary basins within the rift valleys host lakes. The large lakes found in the Albertine Rift, e.g. Lakes Kivu, Malawi and Tanganyika, provide a unique record of climatic changes over the last several million years (Nicholson 1996a, b). The Oldupai and Laetoli Basins of northern Tanzania, however, occur in warps located on the regional plateau to the west of the Gregory Rift.

Fig. 3.2 Schematic diagrams showing a pre-rift stage **a** and a full graben stage **b** with the EARS ascribed to a mantle plume

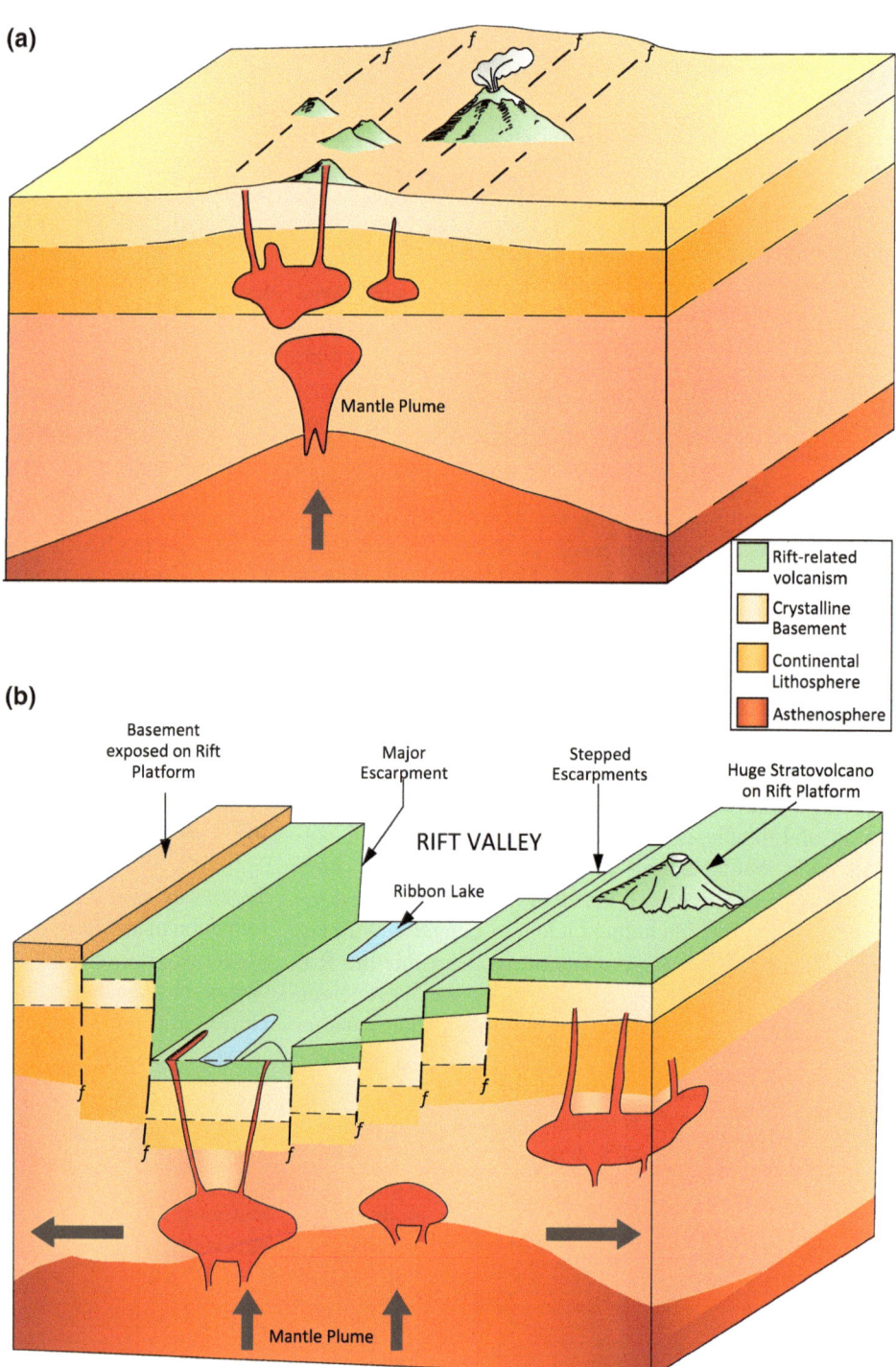

(a)

Mantle Plume

Rift-related
volcanism

Crystalline
Basement

Continental
Lithosphere

Asthenosphere

(b)

Basement
exposed on Rift
Platform

Major
Escarpment

Stepped
Escarpments

Huge Stratovolcano
on Rift Platform

RIFT VALLEY

Ribbon Lake

Mantle Plume

3.8   Lake Victoria and the Victoria Nile

The capture of the upper parts of the Nile River and the creation of Lake Victoria is associated with ongoing rifting during the previous several million years. Lake Victoria with an area of 68,800 km^2 is unusual for the African Great Lakes in that it occurs in a shallow warp located between the Albertine and Gregory Rifts (Fig. 3.1). The relatively shallow depth (<100 m) and rectangular outline contrasts with the deep finger lakes of the Albertine Rift. A maximum age of 1.6 Ma is estimated from sediments exposed on the shores of the Kavirondo Gulf, Kenya (Kent 1944). As rifting progressed, however, the basin tilted eastwards to expose sediments in the Kagera valley, Uganda (at an altitude of 150 m) with an age of 0.8 Ma (Doornkamp and Temple 1966). The lake is underlain by 60-m-thick sediments, consistent with approximately 40,000 years of deposition. Lake

Victoria was initially filled from rivers (headwaters of the Katonga and Kagera) blocked by the rise of the eastern side of the Albertine Rift (Beadle 1981). Renewed episodes of uplift on the western side closed the Beni Gap, such that rivers in the Albertine Rift drained northwards into the Albert Nile (Fig. 3.3). During this phase of activity, the main outflow from Lake Victoria was by the Katonga River into Lakes Edward and George. This system subsequently fed into Lake Albert and the Albert Nile via the Semliki Valley. At this time, the Kafo River flowed westwards directly into Lake Albert.

Development of a gentle upwarp part way between the Albertine Rift and Lake Victoria, at approximately 30,000 BP, resulted in further changes to the drainage patterns of east and central Africa. The connection between Lakes Edward and Victoria was terminated and the River Katonga began flowing eastwards. The rise in water entering Lake Victoria caused a new outlet to form, at Jinja, and the Victoria Nile was borne. The connection of the Victoria Nile

with the Lake Kyoga–River Kafo system is estimated to have occurred at approximately 13,000 BP (Talbot and Williams 2008). Numerous desiccation events, mostly associated with the Late Pleistocene Ice Ages (0.68 Ma–12,000 BP), have been identified in East Africa, as will be discussed further in Chap. 6. During some of these events, Lake Victoria dried entirely. Probably, the most impressive site on the Victoria Nile is the Murchison Falls (Plate 3.2) where 300 m^3/s of water flows over a height of 43 m from the wide river above the falls into a narrow gorge on the eastern edge of the Albertine Rift (Beadle 1981; Roberts 2007). The latter is demarcated by several small faults including the Bunyoro Escarpment. Proximal to the falls the escarpment forms a small step in comparison to the larger escarpment further north. The falls have developed on the contact between resistant granite gneiss of the Uganda Gneiss complex of the regional plateau and the softer rift-related sediments of the Lake Albert basin.

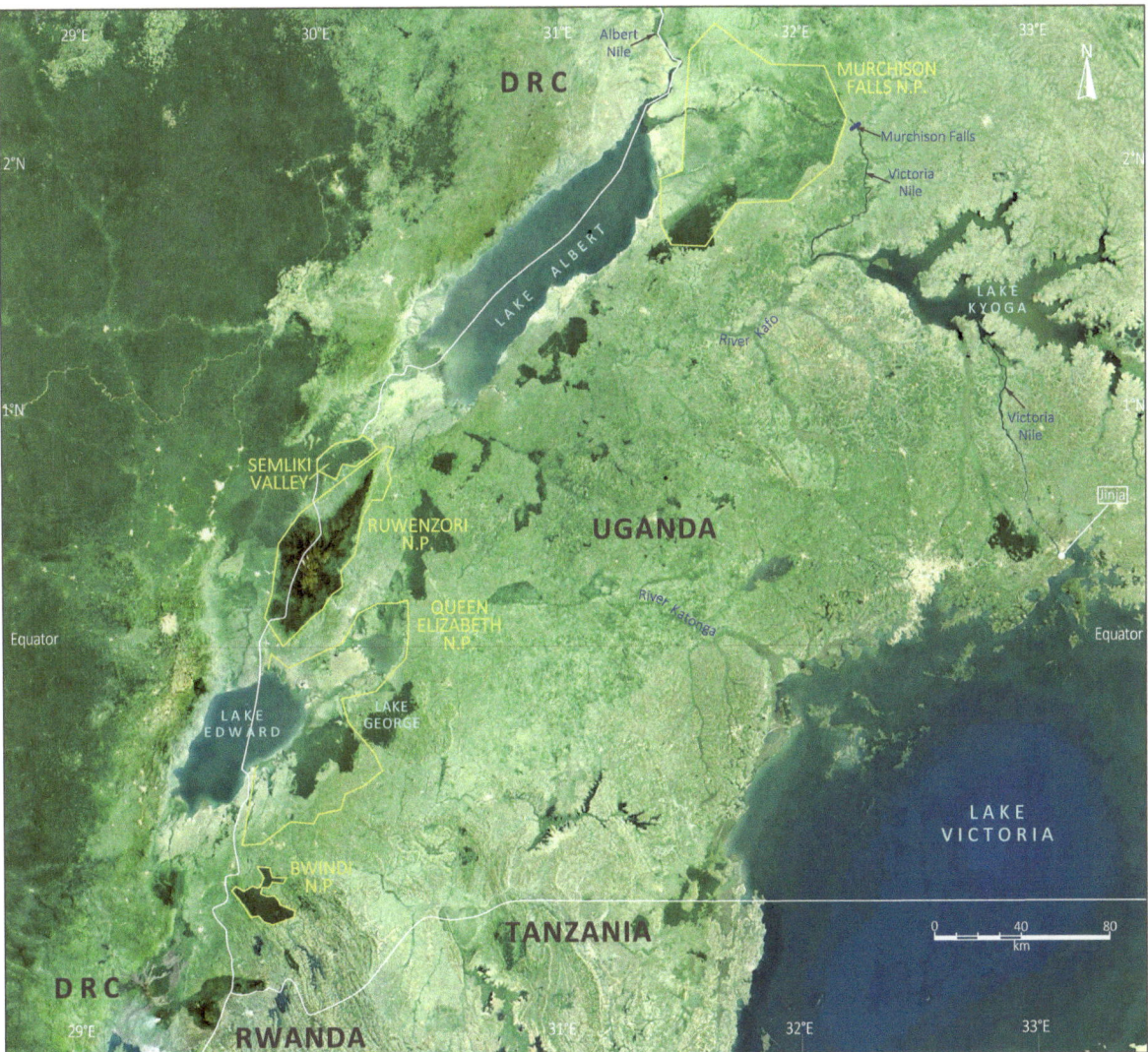

Fig. 3.3 Lakes Albert and Edward in the Albertine Rift demarcate the boundary between Uganda and the DRC. The Victoria Nile exits Lake Victoria at Jinja and flows northwest into Lake Albert via the sinuous Lake Kyoga. *Source* Google Earth Image 2017, Landsat/Copernicus

Plate 3.2 **a** The Victoria Nile plunges over the regional escarpment associated with the eastern edge of the Albertine Rift at Murchison Falls, Uganda; **b** The Victoria Nile below the Murchison Falls is constricted within a narrow gorge comprised of resistant rocks of the Uganda Gneiss complex

Plate 3.3 **a** The Olkaria geothermal field, Hell's Gate National Park is exploiting heat from a relatively shallow magma chamber; **b** Steam is vented from wellheads under excess pressure

References

Bailey, D. K. (1974). Continental rifting and alkaline magmatism. In H. Sorensen (Ed.), *The alkaline rocks* (pp. 148–159). New York: Wiley.

Baker, B. H., Mohr, P. A., & Williams, L. A. J. (1972). Geology of the Eastern Rift System of Africa. *Geological Society of America Special Paper, 136,* 67 p.

Beadle, L. C. (1981). *The inland waters of tropical Africa—An introduction to tropical limnology* (2nd ed., 475 p). London: Longman.

Chorowicz, J. (2005). The East African Rift System. *Journal of African Earth Sciences, 43,* 379–410.

Dawson, J. B. (2008). The Gregory Rift Valley and Neogene-recent volcanoes of northern Tanzania. *Geological Society London Memoir 33,* 102 p.

Doornkamp, J. C., & Temple, P. H. (1966). Surface, drainage and tectonic instability in part of the Southern Uganda. *The Geographical Journal, 132,* 238–252.

Ebinger, C. J., Yemane, T., Harding, D. J., Tesfaye, S., Kelley, S. & Rex, D. C. (2000). Rift deflection, migration, abd propagation: Linkage of the Ethiopian and Eastern rifts, Africa. *Geological Society America Bulletin, 112,* 163–176.

Ebinger, C. (2005). Continental break-up: The East African perspective. *Astronomy and Geophysics, 46*(2), 16–21.

Gregory, J. W. (1896). *The Great Rift Valley* (421 p). London: John Murray.

Kent, P. E. (1944). The Miocene Beds of Kavirondo, Kenya. *Quarterly Journal of the Geological Society London, 100*(1–4), 85–118.

Maasha, N. (1975). The seismicity and tectonics of Uganda. *Tectonophysics, 27,* 381–393.

Merle, G. (2011). A simple continental rift classification. *Tectonophysics, 513,* 88–95.

Morley, C. K., Ngenoh, D. K., & Ego, J. K. (1999). Introduction to the East African Rift System. In C. K. Morley (Ed.), *AAPG Studies in Geology: Vol. 44. Geoscience of Rift systems—Evolution of East Africa* (pp. 1–18).

Nicholson, S. E. (1996a). A review of climate dynamics and climate variability in Eastern Africa. In T. C. Johnson & E. O. Odada (Eds.), *The limnology, climatology and paleoclimatology of the East Africa Lakes* (pp. 25–56). Amsterdam: Gordon and Breach.

Nicholson, S. E. (1996b). Sedimentary processes and signals of past climatic changes in the large lakes of the African Rift Valley. In T. C. Johnson & E. O. Odada (Eds.), *The limnology, climatology and paleoclimatology of the East Africa Lakes* (pp. 367–412). Amsterdam: Gordon and Breach.

Nusbaum, R. L., Girdler, R. W., Heirtzler, J. R., Hutt, D. J., Green, D., Millings, V. E., Schmoll, B.S. & Shapiro, J. (1993). The distribution of earthquakes and volcanoes along the East African Rift system. *Episodes, 16*(4), 16427–16432.

Roberts, A. (2007). *Uganda's great Rift Valley* (211 p). Kampala, Uganda: Graphic Systems.

Rogers, N., MacDonald, R., Fitton, J. G., George, R., Smith, M., & Barreiro, B. (2000). Two mantle plumes beneath the East African Rift System: Sr, Nd, and Pb isotope evidence from Kenya Rift basins. *Earth and Planetary Science Letters, 176*(3–4), 387–400.

Sander, S., & Rosendahl, B. R. (1989). The geometry of rifting in Lake Tanganyika, East Africa. *Journal of African Earth Sciences, 8*(2–4), 323–354.

Sarie, E., Calais, E., Stamps, D. S., Delvaux, D., & Hartnady, C. (2014). Present day kinematics of the East African Rift System. *Journal of Geophysical Research, 119,* 3584–3600.

Stankiewicz, J., & De Wit, M. J. (2006). A proposed drainage model for central Africa—Did the Congo flow eastward. *Journal of African Earth Sciences, 44*(1), 75–84.

Suess, E. (1885–1901). *The Face of the Earth* (Sollas & Sollas, Trans.). Oxford: Clarendon Press (3 volumes).

Talbot, M. R., & Williams, M. A. J. (2008). Cenozoic evolution of the Nile Basin. In H. J. Dumont (Ed.), *The Nile: Origin, environment, limnology and human use* (pp. 37–60). Springer Science-Business Media B.V.

Wilson, J. T. (1973). Continental drift, transcurrent and transform faulting. In A. E. Maxwell (Ed.), *The Sea* (Vol. 4, pp. 623–644). New York:Wiley.

Windley, B. (1977). *The evolving continents* (399 p). New York: Wiley and Sons.

Part II
Regional Geology

Basement Complexes and Regional Plateaus

Abstract

The basement complexes of East Africa are dominated by hard, crystalline, metamorphic rocks from the Archaean and Neoproterozoic Eras. The oldest rocks in this region occur in the Lake Victoria Terrane, a segment of the Central African craton with an age of 2.810–2.560 Ga. The two main components of this terrane are greenstones, a group of metamorphosed, mostly basaltic rocks, and granite-gneiss. The greenstones occur in relatively small, arcuate bodies, with the granite-gneiss forming much larger, overlapping plutons. Sections of the Serengeti Plains to the west of the Gregory Rift Valley in northern Tanzania are underlain by the Lake Victoria Terrane. The Central African craton is bordered on the eastern side by the Mozambique Belt, a Neoproterozoic-age mobile belt which extends from southern Africa into the Arabian Shield. The Mozambique Belt is dominated by quartzite and schist, together with plutons of granite. The Lake Victoria Terrane contains economic mineral deposits, particularly of gold. The Mara gold fields overlap into the Serengeti National Park and were subjected to a 'gold rush' in the 1930s. There are several operating mines near Lake Victoria. The Mozambique Belt contains important gemstone deposits. Locally named minerals such as tanzanite and tsavorite are marketed internationally. The gemstones developed during the high-pressure regimes of regional metamorphism on the edge of the Central African craton. The basement complexes were uplifted and eroded during the breakup of the supercontinent of Gondwana. This process commenced in the Jurassic at approximately 180 Ma and persisted into the Palaeocene. Multiple cycles of erosion during the hot and humid climatic regimes of the Jurassic through Palaeocene planed off the uplifted terranes to create the regional plateaus for which East Africa is so renowned. The most prominent is the 70-Ma-old (Cretaceous) African Erosion Surface which is associated with most of the plateaus at elevations of 1,200–1,500 m.

Keywords

African surface • Basement • Central African craton
Mozambique belt • Regional plateaus • Uplift

Photographs not otherwise referenced are by the author.

© Springer International Publishing AG, part of Springer Nature 2018
R. N. Scoon, *Geology of National Parks of Central/Southern Kenya and Northern Tanzania*, https://doi.org/10.1007/978-3-319-73785-0_4

Plate 4.1 The regional plateau north of the Tarangiri National Park, located southwest of Arusha in northern Tanzania, is underlain by deeply eroded basement complexes with higher relief in the background associated with the Ngorongoro Highlands, part of the EARS

4.1 Introduction

The basement complexes of East Africa are generally restricted to the regional plateau and are rarely observed in the rift valley (Fig. 2.1). Two principal basement terranes are identified in the region, the Archaean–Palaeoproterozoic-age Central African craton and the Neoproterozoic-age Mozambique Mobile Belt (Fig. 2.2). Basement complexes are dominated by hard, crystalline plutonic and metamorphic rocks which typically crop out as small hills or inselbergs. Primary lithologies can be difficult to identify due to their having been subjected to repeated phases of metamorphism and deformation. This is especially true of the metasedimentary and metavolcanic rocks of the Archaean, collectively known as greenstones as the primary lithologies have been metamorphosed to greenschist or amphibolite facies. The granitic plutons within the Archaean are also metamorphosed and deformed but these effects are less severe and the primary compositions and mineralogy are retained, despite the gneissic texture. The younger Mozambique Belt is dominated by quartzite, schist and granite.

Exposures of the basement complexes on the regional plateau are ascribed to near-continental-scale uplift associated with the breakup of the supercontinent of Gondwana. This was triggered by an extreme period of lithospheric thinning and extension that probably affected the entire African Plate. Erosion accompanying uplift produced distinctive land surfaces (Plate 4.1). The paucity of Palaeozoic–Mesozoic rocks in East Africa may be because they were repeatedly eroded during cycles of uplift in the hot and humid climatic regimes that characterised the Mesozoic and Palaeogene.

4.2 Archaean: Lake Victoria Terrane

The Archaean-age cratonic nuclei are the oldest component of the African Plate (Fig. 4.1a). The Tanzania cratonic nucleus constitutes an extensive area of some 500,000 km^2 between the Gregory and Albertine Rifts. Two discrete terranes are identified (Dirks et al. 2015). The oldest, the Dodoma Terrane (3.000–2.900 Ga) is restricted to central and western Tanzania, with the younger Lake Victoria Terrane (2.810–2.560 Ga) cropping out extensively in northern Tanzania, southern Kenya and southeastern Uganda. Both terranes are dominated by large overlapping plutons of granite-gneiss. Greenstones are restricted to smaller arcuate bodies, generally schist belt remnants. The most extensive outcrops of the Lake Victoria Terrane occur within the Serengeti National Park. Four such greenstone xenolithic remnants are identified here, covering an age span of 2.810–2.630 Ga. From oldest to youngest, these are known as Nyanzian, Nzega, Kakamega and Rongo. There is some confusion with terminology as *all* of the greenstones in this area were formerly known as Nyanzian. The plutons of granite-gneiss in the Lake Victoria Terrane were emplaced in several stages (2.720–2.560 Ga). They have been subjected to multiple phases of deformation and metamorphism. The Central African craton is well exposed in sections of the Serengeti National Park where both greenstones, typically in river crossings (Plate 4.2a), and low, whale-backed outcrops of granite-gneiss (Plate 4.2b) are characteristic.

4.3 Palaeoproterozoic: Central African Craton

The Archaean cratonic nuclei were enlarged during the Palaeoproterozoic to form three cratons or microcontinents (Fig. 4.1b). The Central African craton developed from collision and accretion of the Tanzania and Congo nuclei. The enlarged cratons are deep-keeled, stable blocks within the African Plate that are very resistant to younger phases of rifting [e.g. Gondwana related or East African Rift System (EARS)]. Collision zones are preserved in Palaeoproterozoic terranes on the perimeter of cratons: the Central African craton is rimmed to the northwest and southwest, respectively, by the Ubendian and Usagaran complexes (Fig. 2.1). They have an age of 2.100–2.030 Ga (Dirks et al. 2015). The Ubendian complex incorporates the Ruwenzori mountains, an anomalous occurrence of a basement complex located within the Albertine Rift. Palaeoproterozoic terranes are dominated by intensively deformed, metamorphic rocks. Recognition of

the different metamorphic terranes is not straightforward and there is confusion between occurrences of the Usagaran complex and younger Mozambique Belt (Shackleton 1986).

4.4 Mesoproterozoic: Formation of Rodinia

African cratons were enlarged by the development of extensive mobile belts during the Mesoproterozoic, resulting in basement terranes that cover significant areas of the continent (Fig. 4.1c). The Mesoproterozoic reveals evidence of a worldwide series of cratonic collisions associated with repeated cycles of tectonism, metamorphism and magmatic activity. These events culminated in the formation of the supercontinent of Rodinia at approximately 1.100 Ga. Rocks of this age have not, however, been identified within East Africa.

4.5 Neoproterozoic: Breakup of Rodinia

The supercontinent of Rodinia broke up during the Neoproterozoic at approximately 850–700 Ma (Fig. 4.1d). This event overlapped with numerous worldwide cycles of rifting and collision that led to plates reassembling into new supercontinents, such as Pannotia. One such cycle, part of the East African orogeny, was associated with the closure of an ancient ocean between Madagascar and India. This resulted in formation of the Mozambique Belt, a remarkable phenomenon that extends over 5,000 km from Antarctica along the eastern side of Africa into the Middle East (Mosley 1993; Dirks et al. 2015). There is little consensus on the age of these rocks and the maximum age of 800 Ma is an estimate. They may, however, be as young as 500 Ma. The Mozambique Belt is exposed in rivers crossings on the Eastern Serengeti Plains, as well as in the Tsavo West National Park.

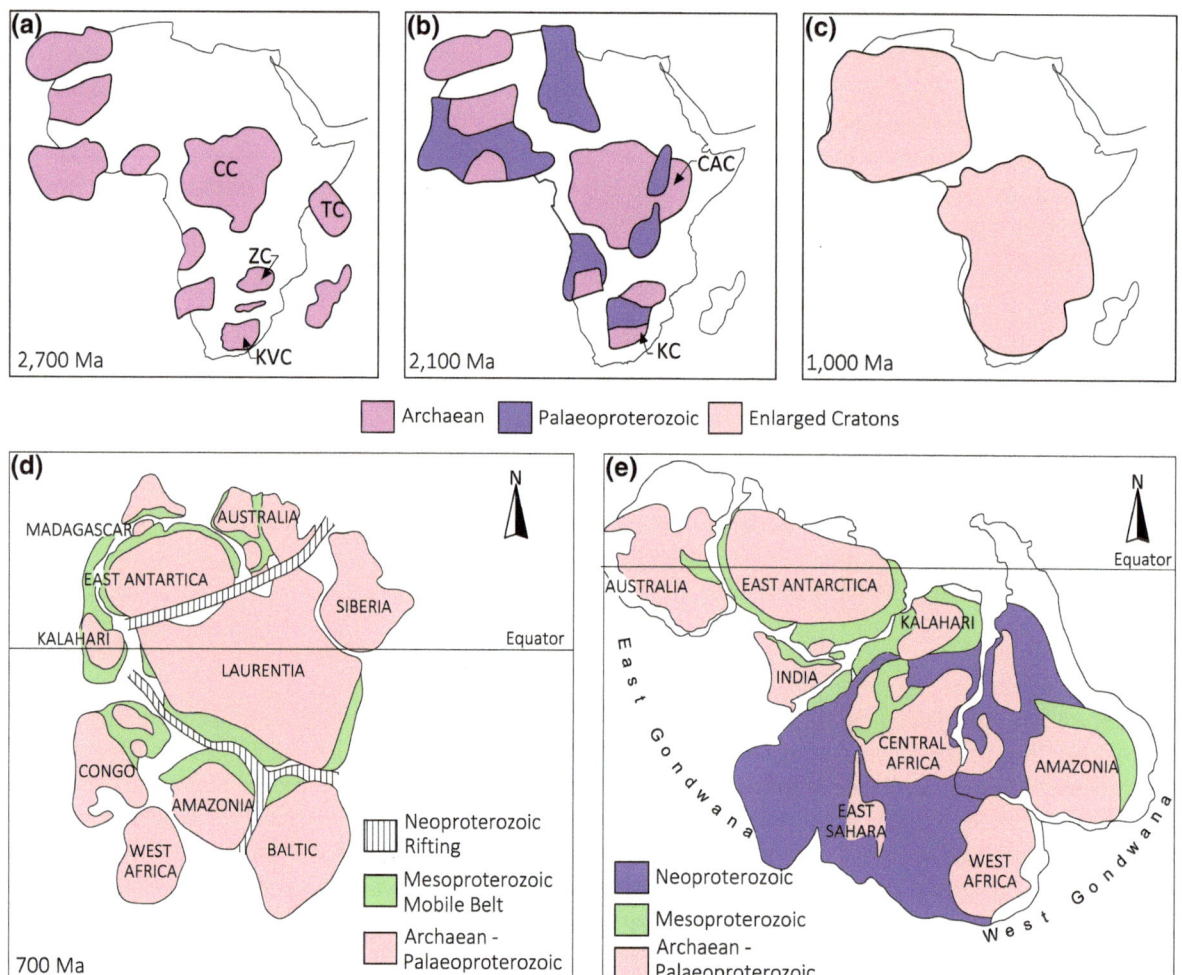

Fig. 4.1 Reconstructions of cratons and supercontinents from an African perspective. **a** The Archaean-age cratonic nuclei—they include the Congo (CC), Kaapvaal (KVC), Tanzania (TC) and Zimbabwe (ZC) components—are not necessarily in their current position; **b** Enlarged Archaean–Palaeoproterozoic cratons include the Central African (CAC) and Kalahari (KC) features (in situ); **c** Cratons were further enlarged into far more extensive basement terranes during the Mesoproterozoic; **d** The supercontinent of Rodinia broke up due to Neoproterozoic-age rifting at approximately 850–700 Ma; **e** Africa existed at the approximate centre of the supercontinent of Gondwana at approximately 250 Ma

4.6 Mesozoic–Palaeogene: Breakup of Gondwana

There is little evidence of the formation of the supercontinent of Pangaea in the Palaeozoic of East Africa. At approximately 250 Ma, however, Africa was probably surrounded by Australia, Madagascar, India and South America, within the newly created supercontinent of Gondwana (Fig. 4.1e). The breakup of Gondwana started in the Jurassic at approximately 180 Ma when rifting occurred on the margins of the African Plate (Begg et al. 2009). A second phase of rifting took place at approximately 150–140 Ma, by which time the margins of the African Plate had thinned considerably. The African Plate probably constituted a discrete entity at this time as both the South Atlantic and Indian Oceans had opened with India, Madagascar and South America having drifted apart. A third major phase of rifting occurred during the Palaeogene at approximately 60–55 Ma, when Australia split from Antarctica. The breakup of Gondwana resulted in some rifting within the African Plate. The northward-flowing section of the Nile River in Sudan and Egypt, for example, is associated with a failed rift that partially developed during the Jurassic–Cretaceous.

4.7 Regional Plateau

The regional plateau that characterises such large parts of the African continent is ascribed to the continental-scale uplift that accompanied the breakup of Gondwana, which in turn was probably triggered by a mantle plume that under-plated the lithosphere (Saggerson and Baker 1965; McCarthy and Rubidge 2005). Repeated cycles of uplift are identified. The oldest occurred during the Jurassic at approximately 200–180 Ma, with younger events having occurred in the Cretaceous and Palaeocene. Each phase of uplift resulted in extensive periods of erosion, most of which occurred during hot and wet climatic cycles. Erosion was so pervasive as to plane off most of the basement complex forming deep, lateritic soil profiles. The most prominent of these is the African Erosion Surface dated at approximately 70 Ma. The regional plateaus with altitudes of 1,200–1,500 m, e.g. parts of the Serengeti Plains and Tsavo West, are associated with this surface. Minor post-African erosion surfaces located at lower elevations are also recognised (dated at approximately 20 and 5 Ma). The regional plateau is enveloped by a coastal escarpment recognisable throughout much of eastern and southern Africa. Erosion has been so extreme that exposures on the regional plateau are typically restricted to inselbergs or isolated koppies, such as the outcrops of granite-gneiss on the Serengeti Plains (Plate 4.3a–b).

4.8 Mineralisation of the Lake Victoria Terrane

The Lake Victoria Terrane hosts the Mara Gold Fields which extend into the Serengeti National Park. There are producing gold mines near Lake Victoria. The area also has potential for nickel deposits and there are numerous occurrences of kimberlite pipes within this area, but few are sufficiently rich in diamonds as to constitute producing mines such as in the vicinity of Shinyanga. The kimberlite pipes, despite being restricted to the Lake Victoria Terrane, are Cretaceous-age intrusives and their formation is believed to be linked to the breakup of Gondwana.

4.9 Gemstones in the Mozambique Belt

This Neoproterozoic mobile belt contains significant deposits of precious and semi-precious gemstones. Most occur within high-grade, granulite facies metamorphic rocks. As many as four phases of deformation may be recognised and temperatures are estimated to have been as high as 1,000 °C with pressures of up to 10–12 kbar. The most well-known gemstone occurrences are in the Merelani region, near Arusha, where deposits of tanzanite are extensively mined. Tanzanite is a unique blue or violet variety of the Ca–Al mineral zoisite, typically found in black shales or schists rich in organic carbon (observed as coarse flakes of graphite) and with anomalous amounts of the trace element vanadium. Tanzanite was discovered by Jumanne Mhero Ngoma in the Merelani Hills, in 1967. Merelaniite is a new mineral discovered in the tanzanite mines. It occurs sporadically as tiny, metallic, dark grey cylindrical whiskers (Jaszczak et al. 2016). A new occurrence of an unusual orange-coloured kyanite has recently been found in the Loliondo area of the Mozambique Belt, to the north of the Ngorongoro Conservation Area (Fig. 1.2). Important occurrences of gemstones also occur in the vicinity of Tsavo East, in southern Kenya. They include gem-quality sapphirine, hibonite and ruby. The gemstones in this area may be associated with small, yet high-grade deposits of graphite. Tsavorite is a gem-quality green-coloured variety of grossularite, a Ca–Al garnet, initially discovered by geologist Geoff Campbell-Bridges at Lemshuko, in northern Tanzania, in 1967, but first mined in 1971 from Tsavo East. The green coloration is ascribed to trace amounts of vanadium or chromium which occur in the host rocks, which are typically graphitic schists that have undergone high-grade metamorphism and intense deformation.

Plate 4.2 a Archaean-age greenstones in the Seronera area of the Serengeti National Park include dark-coloured amphibolite; **b** Archaean-age granite-gneiss in the Lobo area of the Serengeti National Park is characteristically crisscrossed by veins of light-coloured granite pegmatite

Plate 4.3 **a** Archaean-age granite in the Moru Koppies of the Serengeti National Park projects above the regional plateau. Spheroidal boulders form as a result of onionskin weathering (exfoliation) of rocks with very little foliation; **b** Archaean-age granite in the Moru Koppies of the Serengeti National Park Serengeti forms tabular or finger-shaped bodies (also known as tors) due to localised jointing

References

Begg, G. C., Griffin, W. L., Natapov, L. M., O'Reilly, S. Y., Grand, S. P., O'Neill, C. J., et al. (2009). The lithospheric architecture of Africa: Seismic tomography, mantle petrology, and tectonic evolution. *Geoscience, 5,* 23–50.

Dirks, P. H. G. M., Blenkinsop, T. G., & Jelsma, H. A. (2015). The geological evolution of Africa. In *Geology volume IV* (15 p). Oxford: Encyclopedia of Life Support Systems (EOLSS).

Jaszczak, J. A., Rumsey, M. S., Bindi, L., Hackney, S. A., Wise, M. A., Stanley, C. J., et al. (2016). Merelaniite, $Mo_4Pb_4VSbS_{15}$, a New Molybdenum-Essential Member of the Cylindrite Group, from the Merelani Tanzanite Deposit, Lelatema Mountains, Manyara Region, Tanzania. *Minerals, 6*(4), 115–119.

McCarthy, T. S., & Rubidge, B. (2005). *The Story of Earth and Life: A southern Africa perspective on a 4.6 billion year journey* (334 p). Cape Town: Struik.

Mosley, P. P. (1993). Geological evolution of the late proterozoic "Mozambique Belt" of Kenya. *Tectonophysics, 221,* 223–250.

Saggerson, E. P., & Baker, B. H. (1965). Post-Jurassic erosion surfaces of East Africa and their deformation in relation to rift structure. *Journal of the Geological Society of London, 121,* 51–72.

Shackleton, R. M. (1986). Precambrian collision tectonics in Africa. In M. P. Coward & A. C. Ries (Eds.), *Collision tectonics* (Special Publication Vol. 19, pp. 324–349). Geological Society London.

The Gregory Rift

5

Abstract

The three stages of development of the East African Rift System (EARS), pre-rift, half-graben and full graben, can be related to specific episodes of faulting and volcanism within the Gregory Rift. The pre-rift stage (Late Oligocene–Miocene) initiated formation of the regionally extensive Kenya-Tanzanian Dome, with plateau-style volcanic outpourings and isolated shield volcanoes, e.g. Mount Elgon. Formation of the narrow rift valley in central Kenya commenced during the half-graben stage (Late Miocene–Pliocene) and persisted, particularly in southern Kenya into the full graben stage (Pleistocene). These events triggered extensive plateau-style volcanism which shows a pattern of younging southwards and with the youngest sequences located in the centre of the valley, i.e. as lateral extension of the rift persisted. These events were also accompanied by eruption of discrete volcanic cones throughout central/southern Kenya, albeit they reveal a more chaotic distribution relative to their ages. The extensiveness of volcanism on the Eastern Rift Platform is a notable feature of the Gregory Rift in central/southern Kenya, e.g. Aberdare Range (Miocene), Mount Kenya (Pliocene) and Chyulu Hills (Pleistocene–

Holocene). Rifting in northern Tanzania is restricted to the half-graben stage (Pliocene–Pleistocene), an observation consistent with the southward propagation of this branch of the EARS. The rift diverges into three arms within a 200-km-wide, structurally complex area with extensive volcanism. Two of the rifts, including the Natron–Manyara half-graben are a continuation of the full graben in southern Kenya; they peter out southwards. The volcanism of northern Tanzania is divided into an older group (Pliocene–Early Pleistocene), of which the Ngorongoro Volcanic complex is the most well-known example, and a younger group, the latter including the dormant Kibo component of the multicentred Kilimanjaro edifice and the active cones of Mount Meru and Oldoinyo Lengai. Despite the Gregory Rift being dominated by volcanic rocks, localised sedimentary basins occur in the rift valley, e.g. the Lake Manyara basin, and in small warps on the rift platforms, e.g. the Oldupai and Laetoli basins.

Keywords

Caldera • Graben • Plateau-style volcanism
Rifting • Sedimentary basin • Volcanic cone

Photographs not otherwise referenced are by the author.

© Springer International Publishing AG, part of Springer Nature 2018
R. N. Scoon, *Geology of National Parks of Central/Southern Kenya and Northern Tanzania*, https://doi.org/10.1007/978-3-319-73785-0_5

Plate 5.1 The escarpment west of Lake Manyara, the western boundary fault of the Natron–Manyara half-graben, consists of multiple layers of volcanic lavas, ashes and pyroclastics

5.1   Introduction

The Gondwana-related cycles of uplift and erosion that occurred during the Jurassic, Cretaceous and Palaeogene periods, resulting in formation of the regional plateau, as described earlier, were followed in East Africa by a renewed period of lithospheric thinning that ushered in the onset of the EARS. The boundaries of the African Plate have changed little in the past 200 million years and the faulting and volcanism of the EARS may be ascribed to rejuvenation of the Gondwana-related mantle plume (Chorowicz 2005). The extensiveness and chemical features of the volcanism in Kenya led Rogers et al. (2000) to postulate the existence of two mantle plumes beneath the EARS. The Ethiopian, Albertine and Gregory Rifts each have unique features (Chap. 3).

The three stages of rifting that characterise continental rifts, i.e. pre-rift, half-graben and full graben, are consistent with the southward propagation of the Gregory Rift. This rift is typically defined by zones of normal faulting (Baker and Wohlenberg 1971; Baker et al. 1972) that have created

regional escarpments, as seen on a digital terrane image (Fig. 5.1).

Escarpments typically consist of multiple, near-horizontal layers of volcanic lavas, ashes and pyroclastic rocks (Plate 5.1). The larger faults are accompanied by subsidiary faults that create a stepped topography (Plate 3.1). The platforms that border the Gregory Rift, particularly in central/southern Kenya occur at considerably higher elevations than the Gondwana-age plateau, a well-known example being the Eldoret Plateau (Fig. 1.2), the home of many of East Africa's most famous long-distance athletes.

The Gregory Rift has an extremely high volcanic output, particularly in comparison with the Albertine Rift. Some sections of the rift valley are wholly volcanic. The intensity of the volcanism is manifested by both plateau-style outpourings and giant edifices (Baker 1987; Dawson 2008). The composition of the volcanic rocks in the Gregory Rift has been widely studied. Three groups have been identified. Each group has a unique fractionation trend that can be related to the three stages of rifting.

Sedimentary basins associated with the Gregory Rift occur both within the rift valley and in warps and flexures on

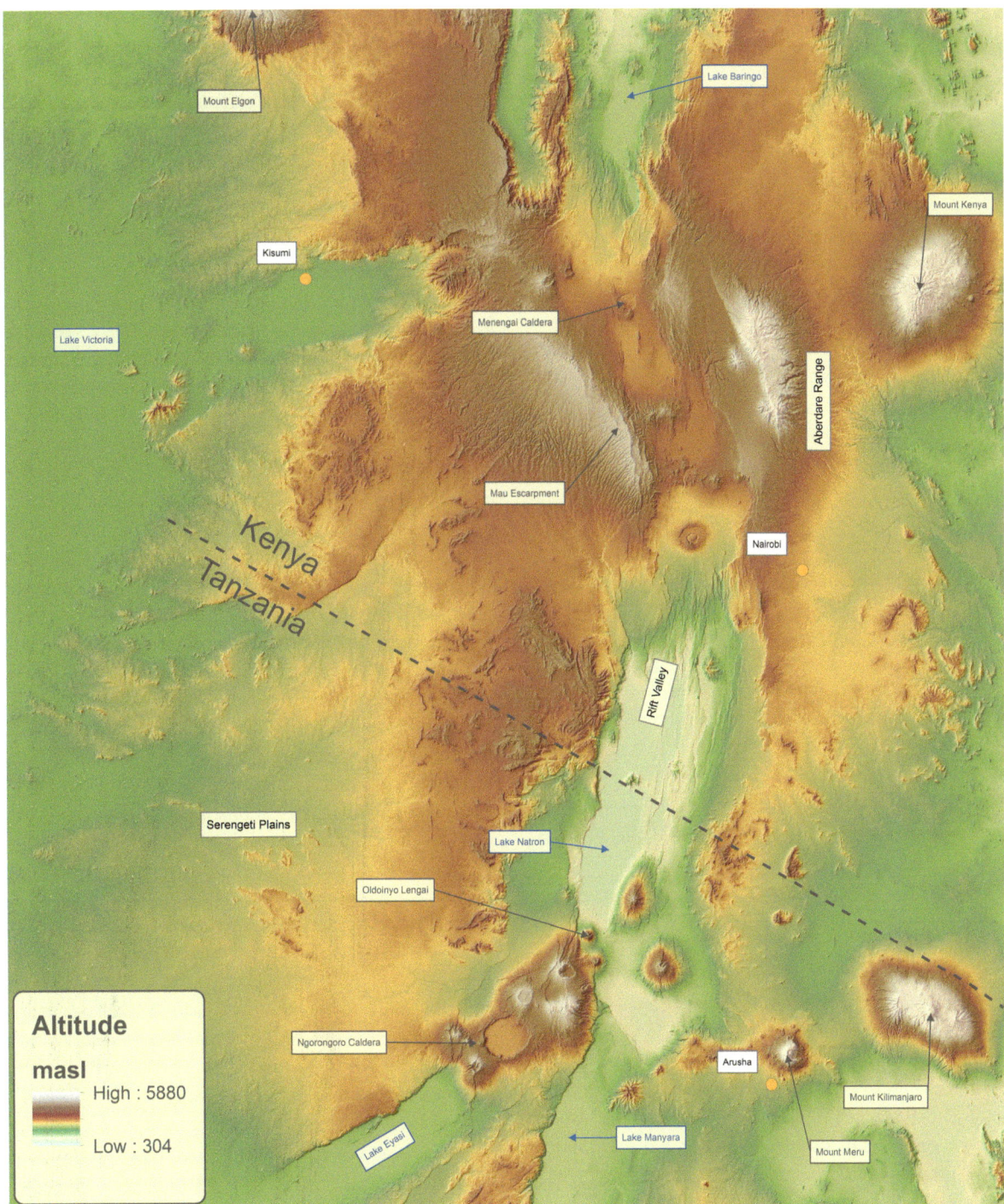

Fig. 5.1 A digital terrane image shows the principal physiological features of the region, the narrow Gregory Rift Valley in central and southern Kenya, the divergence in northern Tanzania, the high altitude of the rift platforms and the immensity of some of the volcanic edifices. Courtesy of Oliver Burdekin, birdGIS, with digital terrain shading created using ArcGIS® software by Esri

some of the rift platforms. Most of the ribbon lakes are shallow and alkaline, indicative of the narrowness of the valley and the restricted catchment.

5.2   Three Stages of Rifting

The Gregory Rift propagated southwards into northern Kenya, probably as an extension of the Early Oligocene-age rifting in Ethiopia. The migration southwards, initially into central/southern Kenya and subsequently into northern Tanzania, is consistent with the following three stages recognised in each section:

- Central/southern Kenya (after Baker et al. 1972):
    - Pre-rift stage (including regional doming): Late Oligocene–Early Miocene (22–12 Ma);
    - Half-graben stage: Late Miocene–Early Pliocene (12–4 Ma) and
    - Full graben stage: Pliocene (4–1.8 Ma), peaking in the Pleistocene at 1.5 Ma, and persisting into the Holocene.

- Northern Tanzania (after Dawson 2008):
    - Pre-rift stage (including regional doming): Late Miocene (?);
    - Half-graben stage: Pliocene and Pleistocene (4–1.0 Ma) and persisting into the Holocene; and
    - Full graben stage: poorly developed (Pangani Rift?).

5.3   Subdivision of the Volcanism

The pre-rift volcanism of central Kenya (Late Oligocene–Early Miocene) is dominated by plateau-style volcanic outpourings; there are few discrete cones, with the notable exception of Mount Elgon. The half-graben and full graben stages of rifting in central/southern Kenya are consistent with a general trend of the plateau-style outpourings becoming progressively younger southwards (Baker 1987): Late Miocene in the northern part of central Kenya, Pliocene in the southern part of central Kenya, and Pleistocene-Holocene in southern Kenya. The ages of the discrete volcanic cones in Kenya, however, reveal a chaotic distribution.

In northern Tanzania, the pre-rift (Late Miocene) volcanism is probably obscured in faulted volcanic terranes on which volcanism associated with the half-graben stage of rifting developed. Two groups of volcanism, encompassing both the plateau-style outpourings and discrete cones, are recognised [Dawson (2008) and earlier contributions]:

Pliocene–Early Pleistocene ('Older Volcanism') and Late Pleistocene–Holocene ('Younger Volcanism').

5.4   Pre-rift Stage of Doming and Volcanism in Central/Southern Kenya

The pre-rift stage of the EARS was manifested by the onset of regional doming and volcanism. The Late Oligocene–Early Miocene volcanic terranes include extensive plateau-style outpourings that flooded and smoothed out the irregularities of palaeo-landscapes. This resulted in flat-topped or stepped features, some of which are preserved on the rift platforms (Figs. 5.2 and 5.3). Examples of this early volcanism include the Nairobi Volcanic Field (13.4 Ma) and the Yatta Plateau (13.2 Ma), the latter being a 20-m-thick feature that extends unbroken for some 300 km in southeastern Kenya. The Yatta Plateau may be the most extensive lava flow ever recorded, although there is some debate and it may be a series of multiple flows. The oldest of the discrete cones in Kenya is probably the Kisingiri Volcano (38–17.4 Ma) (Woolley 2001). The largest and most well known of the pre-rift volcanic cones associated with the Gregory Rift is, however, Mount Elgon (22 Ma), which straddles the border between Kenya and Uganda. Large sections of the pre-rift volcanism of central/southern Kenya are, however, obscured by younger sequences. Volcanism was more significant than faulting, with the first evidence (Early Miocene) of development of a rift valley being a chain of marginally warped depressions.

5.5   The Half-Graben and Full Graben Stages in Central/Southern Kenya

The rift valley to the south of Lake Turkana constitutes a relatively narrow (40–65 km wide) feature defined by large, normal faults (Figs. 5.2 and 5.3). The larger faults are generally expressed on surface as regional escarpments. Two offshoots of the Gregory Rift are recognised, the Kisumu Rift, which includes an arm of Lake Victoria, and the Isuria Rift, which extends into the Serengeti Plains. The half-graben stage commenced in this region with asymmetrical faulting of the earlier-formed depressions (Baker et al. 1972). This process commenced in the Late Miocene and persisted into the Early Pliocene. This Late Miocene–Early Pliocene-age faulting is preserved in some boundary faults, e.g. the Elgeyo Escarpment. Sections of the rift valley that are broader and poorly defined, e.g. in the vicinity of Lakes Baringo and Bogoria, represent areas of the half-graben stage.

The full graben stage of rifting occurs in sections of central/southern Kenya where the rift valley is particularly

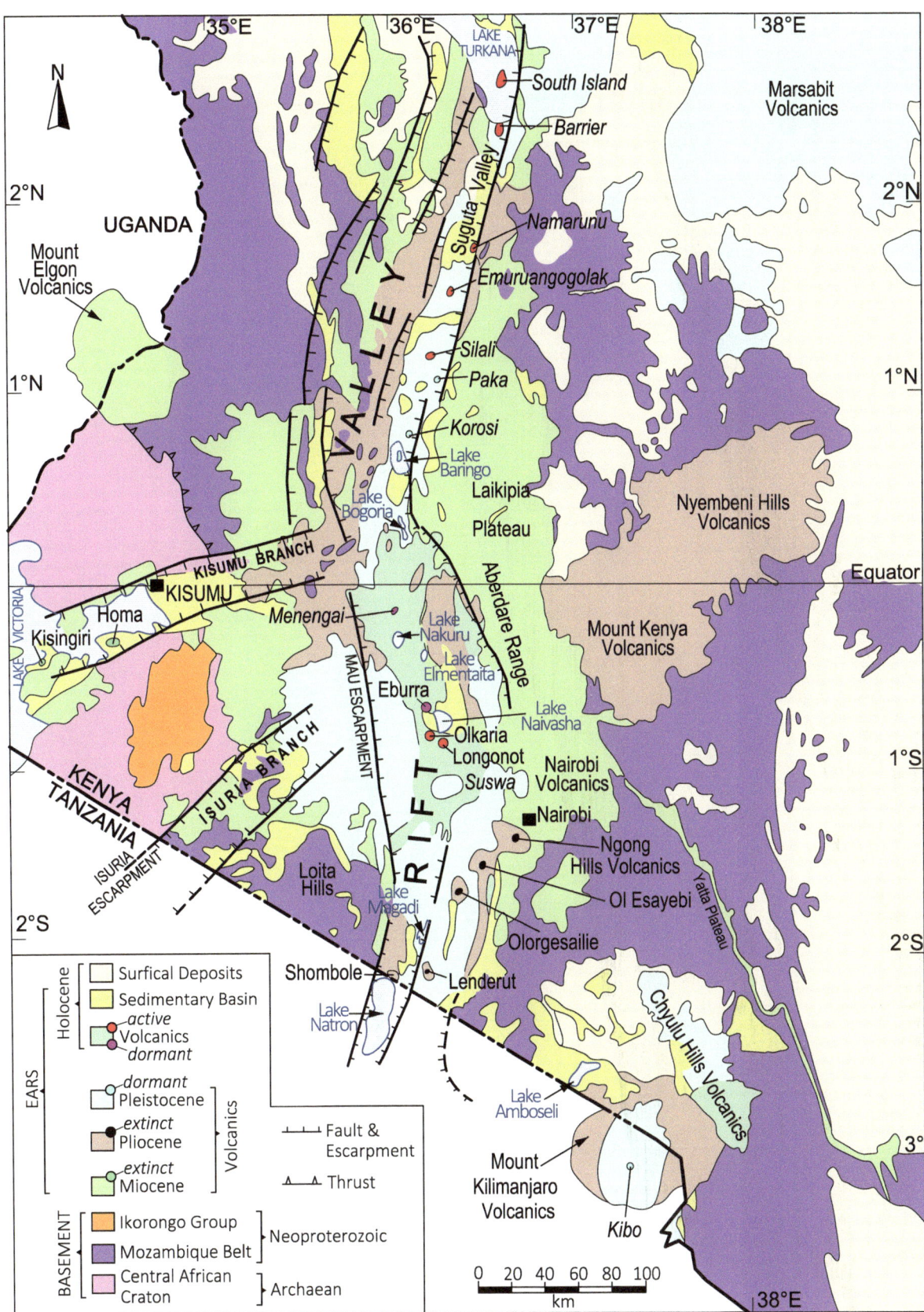

Fig. 5.2 Simplified geological map of central and southern Kenya compiled from various sources including the 1:1,000,000 scale map of the Ministry of Energy and Regional Development of Kenya (1987 Edition) and with some updates and changes. The Miocene-age volcanism is mostly associated with the half-graben stage of rifting (Late Miocene), as the pre-rift (Early Miocene) volcanism is of restricted occurrence

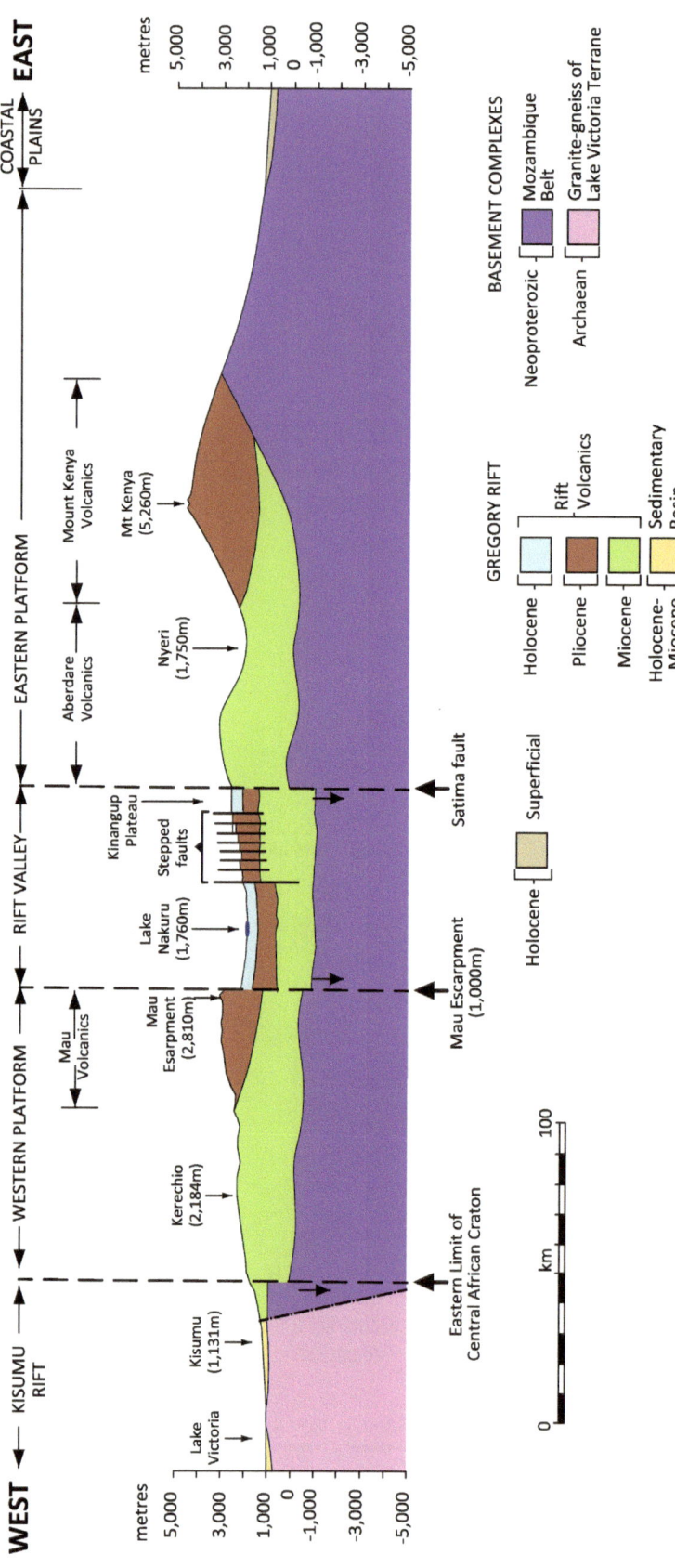

Fig. 5.3 Geological section of central Kenya based on Lake Nakuru (latitude 0° 21' South) and extending from the Kisumu Gulf (Lake Victoria) to the eastern slopes of Mount Kenya. The thickness of the volcanics associated with the Gregory Rift is schematic. The youngest (Holocene) volcanics are restricted to the centre of the rift valley with the Older (Miocene and Pliocene) Volcanics on the rift platforms. The Mount Kenya Volcanics are an example of the extensiveness of the volcanism on the Eastern Rift Platform. Pleistocene-age volcanics are missing from this section but they can be assumed to occur in stratigraphic sequence

well defined, notably between Lakes Nakuru and Magadi (Fig. 5.2). Prominent boundary faults occur on both the western and eastern sides of the rift valley, e.g. the Mau Escarpment (west) and the Satima Fault (east) in the vicinity of Lake Nakuru. The peak of the tectonism to date occurred in the Early Pleistocene (Baker et al. 1972). The Late Pleistocene experienced a rejuvenation but faulting was less intense. Volcanic cones and calderas are prominent in the full graben section of the rift valley, i.e. Menengai Caldera, Mount Longonot and Suswa Volcano. The Late Pleistocene faulting also produced numerous small step faults, e.g. in the area east of Lake Nakuru. Some of these features can be identified in the cross section (Fig. 5.3).

The majority of faults associated with the Gregory Rift are orientated approximately north–south, consistent with the regional nature of the EARS. They typically extend on surface for a maximum of 20–25 km. The outline of rifts on regional images (e.g. Fig. 3.1) is very schematic. Faults orientated north–northeast are indicative of localised reactivation of northeast trending Gondwana-age structures (Chorowicz 2005). Some of the escarpments associated with the rift valley are massive features. Examples include the

Losiolo Escarpment (2,000 m) on the eastern side of the Suguta Valley and the Mau Escarpment (1,000 m) west of Lake Nakuru (Frontispiece 2). The Nguruman Escarpment near Lake Magadi separates the basement terranes of the Loita Hills from the rift valley.

5.6   The Northern Tanzania Divergence

The northern Tanzania divergence is a 200-km-wide, structurally complex section of the Gregory Rift (Dawson 2008; Le Gall et al. 2008). Three branches or arms are identified (Fig. 5.4). Two of the arms, Natron–Manyara and Eyasi, are half-grabens that peter out southwards, but the Pangani Rift is a full graben that extends into the Indian Ocean and may connect with the mid-Ocean ridge. Despite the Natron–Manyara half-graben being a continuation of the full graben in southern Kenya, it is bordered to the east by a gentle, westward-sloping plain that extends as far as Moshi. There is little evidence of an eastern boundary fault, although the Western Escarpment forms an impressive barrier (Frontispiece 1) that connects northwards with the Nguruman Escarpment and persists southwards for several tens of

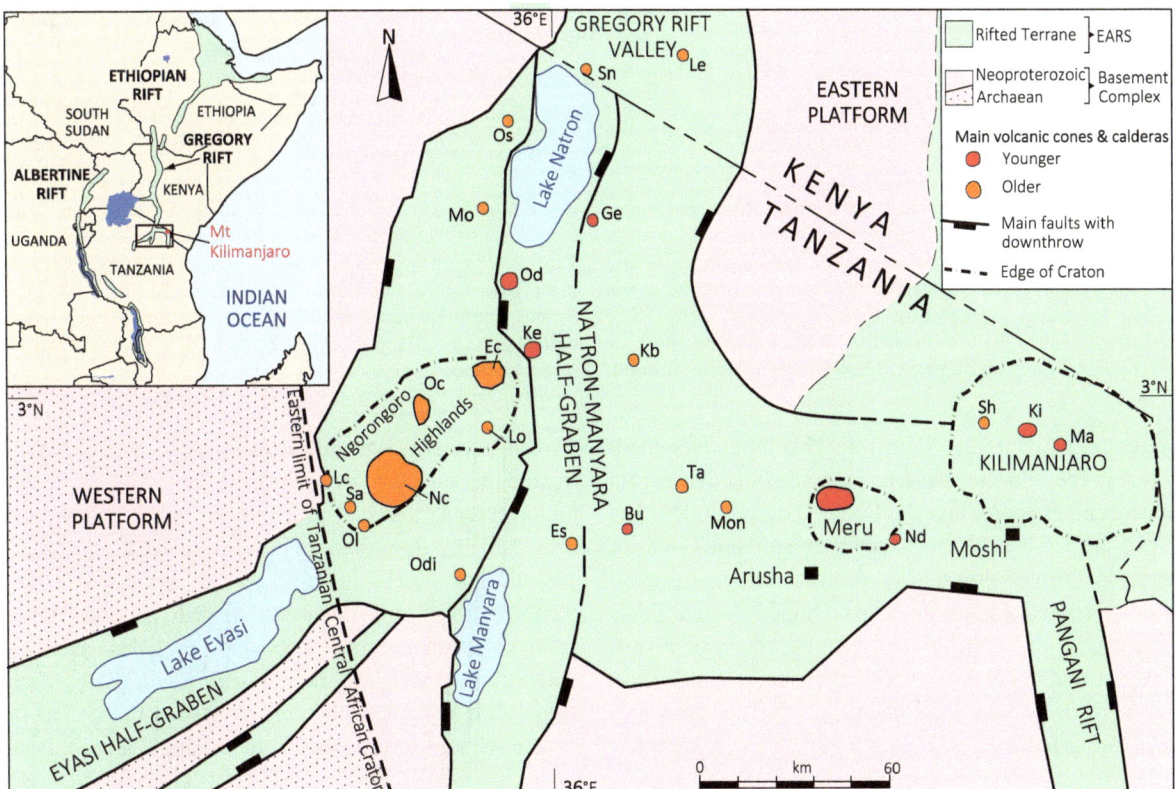

Fig. 5.4 Simplified structure of the northern Tanzanian divergence based on Dawson (2008). Rifts are mostly infilled by volcanic terranes with some small sedimentary basins (not shown). The Tanzanian Craton, part of the Central African craton occurs to the west of the dashed line; the Mozambique Belt occurs to the east

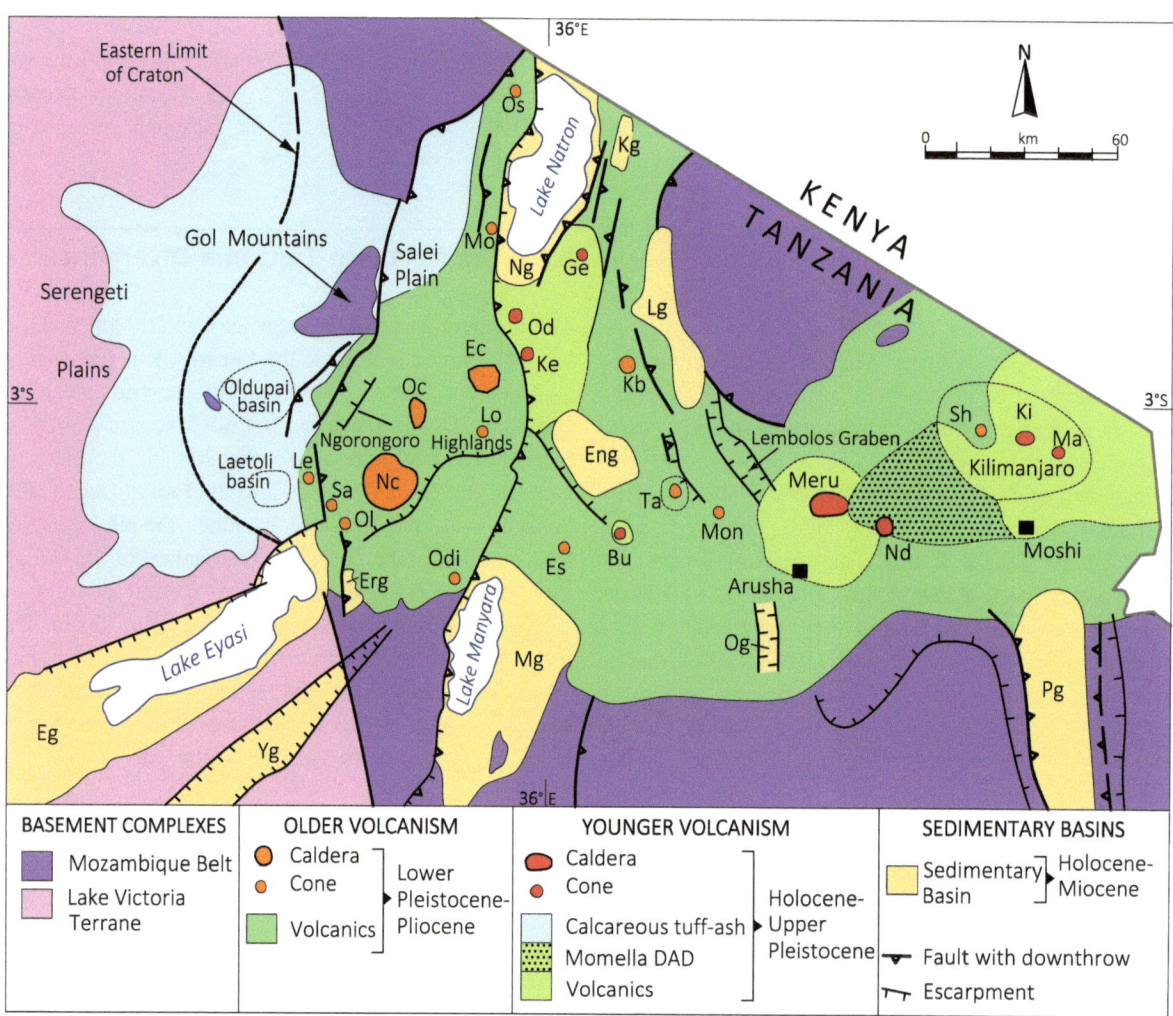

Fig. 5.5 Simplified geological map of northern Tanzania (after Dawson 2008). The divergence includes small grabens, but is dominated by plateau-style volcanics, the Ngorongoro Volcanic complex and numerous additional volcanic edifices. Discrete cones include the following: Burko (Bu); Essimingor (Es); Gelai (Ge); Kerimasi (Ke); Kibo (Ki); Ketumbeine (Kb); Lemagrut (Le) or Makarot; Loolmalasin (Lo); Mawenzi (Ma); Monduli (Mon); Mosonik (Mo); Ngurdoto (Nd); Oldeani (Ol); Oldoinyo Dili (Odi); Oldoinyo Lengai (Od); Oldoinyo Sambu (Os); Sadiman (Sa); Shira (Sh) and Tarosera (Ta). Three of the cones include giant calderas: Empakaai (Ec); Ngorongoro (Nc) and Olmoti (Oc). Sediment-filled grabens include the following: Engaruka (Eng); Erumkoko (Erg); Eyasi (Eg); Lembolos (Lg); Manyara (Mg); Natron (Ng); Oljoro (Og); Pangani (Pg) and Yaida (Yg)

kilometres west of Lake Manyara. This boundary fault developed in the Pliocene and was particularly active during the Pleistocene at approximately 1 Ma (Dawson 2008). The Eyasi half-graben can be envisaged as a structural step, or platform, on the western side of the rift valley, separated from the Serengeti Plains by several modest-sized faults downthrown to the east. In the vicinity of Lake Eyasi, however, an additional huge escarpment has developed in association with a major fault. The Eyasi half-graben contains the Ngorongoro Highlands, a major volcanic terrane.

The northern Tanzania divergence is a complex, rifted terrane that includes, in addition to the features noted above, minor grabens and half-grabens, e.g. the Lembolos Graben (Fig. 5.5), as described by Dawson (2008). Some of the downthrown blocks in this area are associated with sedimentary basins, e.g. Engaruka, Lake Eyasi and Lake Manyara. The divergence probably includes plateau-style outpourings related to the pre-rift stage (Late Miocene), although few of these features have been dated. The deeply eroded cone (thus assumed to be relatively old) of the

Oldoinyo Dili Volcano is an anomalous feature which may also be part of the pre-rift stage. The divergence is, however, mostly known for the abundance of giant volcanic edifices that are Plio-Pleistocene in age or younger, as described below.

The northern Tanzania divergence probably formed in response to impact between the southward-propagating rifting and the Central African craton (Ebinger et al. 1997), a deep-rooted block of ancient crystalline rocks that is massively resistant to deformation. The rift can be envisaged as having refracted eastward away from the craton. In comparison, the Neoproterozoic Mozambique Belt is more readily stretched and thinned. A section based on the Ngorongoro Caldera illustrates the principal features of this area (Fig. 5.6).

5.7 Volcanism of the Half-Graben and Full Graben Stages in Central/Southern Kenya

There is considerable overlap between the three ages of rift-related volcanism shown on the map (Fig. 5.2), but two broad relationships are emphasised. First, the volcanism gets younger southwards. Second, the Miocene and Pliocene volcanism is mostly exposed on the rift platforms, with the more recent activity occurring in the rift valley.

The **Late Miocene** volcanism is particularly extensive on the Eastern Rift Platform. Examples include the Laikipia Plateau (undated), the Aberdare Range (undated) and the Maralal Volcanics (the latter persisted into the Pleistocene). The Mau Plateau on the Western Rift Platform also includes large areas of Late Miocene-age volcanics. Discrete cones of this age include the long-lived Homa Volcano (12–1.3 Ma), near Lake Victoria and the Ol Esayeiti Volcano (6.7–3.6 Ma) near Nairobi. Most of the individual volcanic centres have been reviewed by Woolley (2001) who also presents the available radiometric-age dates.

Pliocene-age volcanic terranes are prevalent, both on the Western Rift Platform between Lakes Turkana and Nakuru, and on the Eastern Rift Platform to the north and south of Nairobi. The latter area includes the Nyambeni Hills Volcanics (4.5–0.46 Ma) and the Mount Kenya Volcanics (4.5–2.64 Ma). The Ngong Hills, an area made famous by Karen Blixen's book 'Out of Africa', includes four peaks (maximum altitude of 2,460 m) which are the remnants of a volcano (2.58 Ma) with an estimated diameter of 11 km. The Namarunu Volcano in the Suguta Valley is unusual in that it dates from the Late Miocene with volcanism persisting into the Pleistocene (6.8–0.5 Ma). There are several discrete cones in the southern part of the rift valley, e.g. Olorgesailie (2.62–2.21 Ma), Lenderut (2.62–2.53 Ma) and Shombole (2.0–1.96 Ma), the latter being a large, deeply eroded cone with a diameter of 9 km that rises to a height of 1,570 m. These latter cones are unusual as they are older (Pliocene) than the plateau-style outpourings (Pleistocene).

Pleistocene-age volcanism is most prevalent in the southern part of the rift valley, e.g. the Lake Magadi Volcanics (1.42–0.63 Ma). A number of discrete cones became active at this time; the oldest ages are recorded here (Holocene activity is described in Chap. 6). From north to south they include the following (Fig. 5.2). The Barrier Volcano (1.37 Ma) forms a natural partition at the southern end of Lake Turkana with a base diameter of 20 by 15 km. The Emuruangogolak Volcano (0.9 Ma) occurs in the Suguta Valley, as does the giant Silali Volcano (0.28 Ma) which crops out over an area of 30 by 25 km and towers some 760 m above the valley floor. Silali includes a large (7.5 by 5 km), mostly intact caldera. Paka (0.39–0.10 Ma) and Korosi (0.38–0.10 Ma) are large shield volcanoes located to the north of Lake Baringo. The Menengai Volcano (0.18 Ma) includes one of the world's largest intact calderas with an internal area of 12 by 8 km. The huge cone of the Suswa Volcano (0.24–0.10 Ma) dominates the valley to the south of Lake Naivasha. Large volcanic fields on the Eastern Rift Platform of this age include the Marsabit Volcanics (0.6–0.2 Ma) and the northwestern component of the Chyulu Hills (oldest age recorded is 0.48 Ma).

5.8 Rift-Related Volcanism in Northern Tanzania

Most of the prominent volcanic cones and giant edifices of this region have been described by Woolley (2001) and Dawson (2008), including references to the radiometric-age dates quoted herein (which include data of Evans et al. 1971). Many are associated with discrete peaks or mountainous areas visible on a digital terrane image (Fig. 5.7). The principal expression of the **Older Volcanism** (Pliocene–Early Pleistocene) is the Ngorongoro Volcanic complex (4.9–1.8 Ma), a group of eight overlapping volcanoes, some

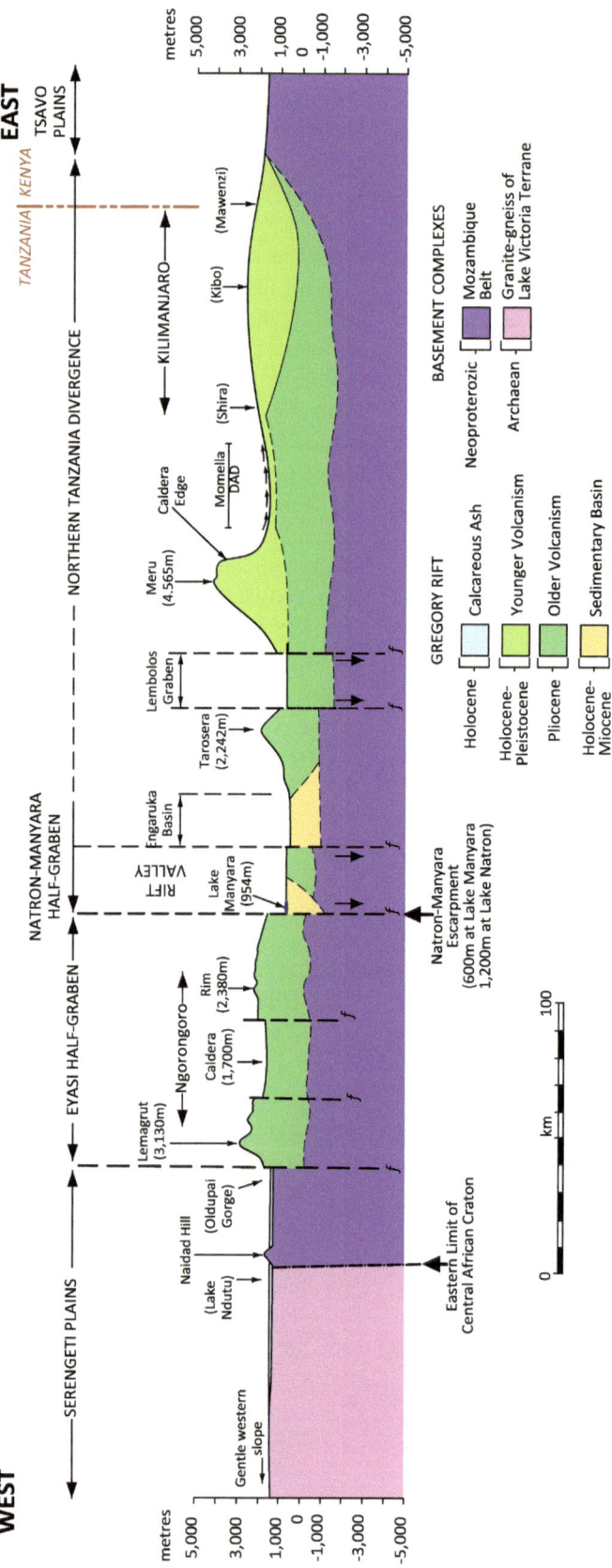

Fig. 5.6 Geological section of northern Tanzania based on the Ngorongoro Caldera (latitude 3° 10' South) and extending from the Eastern Serengeti Plains to Mount Meru and the lower, southern slopes of Kilimanjaro. Localities in brackets (Lake Ndutu, Oldupai Gorge, Kibo, etc.) are located north of the line but are included for reference. The thickness of the volcanics associated with the Gregory Rift is schematic. The Older Volcanism (mostly Pliocene on this section) is particularly extensive in the Eyasi half-graben with the Younger Volcanism restricted to discrete centres in the divergence, notably Meru and parts of Kilimanjaro.

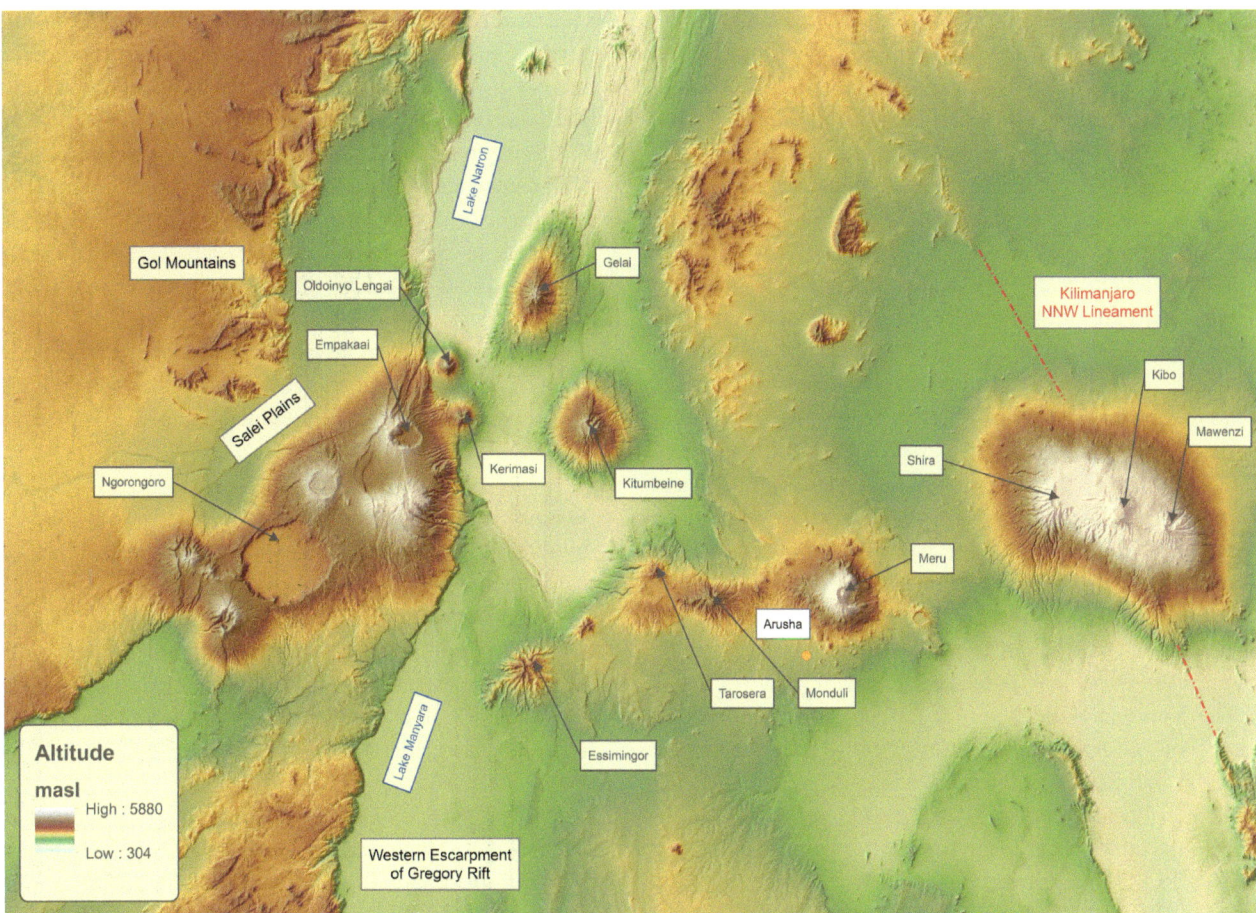

Fig. 5.7 A digital terrane image of part of northern Tanzania shows the prominent escarpments on the western sides of the Natron–Manyara and Eyasi half-grabens as well as the larger volcanic cones and calderas. Courtesy of Oliver Burdekin, birdGIS, with digital terrain shading created using ArcGIS® software by Esri

with giant calderas (Fig. 5.5). The roots to this complex are exposed at the base of the Western Escarpment, notably in the Engare Sero Gorge near Lake Natron. Basaltic and nephelinite lavas and ashes can be observed together with debris flow deposits (Plates 5.2, 5.3, and 5.4). Several discrete cones are also located within the Eyasi half-graben, including the 2,043-m-high Oldoinyo Samba Volcano (2.02 Ma).

Cones located in the faulted terrane of the divergence include Tarosero (2.4–1.9 Ma), a gently sloping, deeply eroded feature, Monduli (2.15–2.09 Ma), and the extensive, dome-like mass of Ketumbeine (1.87–1.54 Ma) which peaks at 2,658 m. The Essimingor Volcano (2,165 m) is the oldest of this group (6.0–3.2 Ma), as described by Mana et al. (2012). Many of these lesser known, yet giant volcanic domes, are protected in parks as they host extensive montane forests. The Kilimanjaro edifice, which is comprised of three discrete cones, is also located in the divergence. Shira (2.5–1.9 Ma) is the oldest component. The Hanang Volcano is the most southerly of the cones in this area, with an age that overlaps the Older and Younger Volcanism (1.5–0.9 Ma).

The **Younger Volcanism** of northern Tanzania is comprised of a half-dozen major cones (Fig. 5.5). The 1,702-m-high, dome-like mass of Mosonik (possible age of 1.28 Ma) is located above the escarpment near Lake Natron. Three large cones occur at the base of this escarpment, Gelai (0.99–0.96 Ma), Kerimasi (0.6–0.4 Ma) and Oldoinyo Lengai (0.05 Ma). The two highest peaks of Kilimanjaro, Mawenzi (1.0–0.45 Ma) and Kibo (0.4 Ma), are part of the Younger Volcanism, as is Burko (1.03–0.97 Ma) and Mount Meru (0.4 Ma). Some of the younger volcanoes are active (Chap. 6).

5.9 Composition of the Volcanism

Continental rifts are invariably associated with the relatively uncommon alkaline magmas (Bailey 1974; Barberi et al. 1982). They are defined as those with an excess of the alkali metals (Na_2O and K_2O) relative to silica (SiO_2), as illustrated on a chemical plot known as a TAS diagram (Fig. 5.8a). Some alkaline rocks are so enriched in Na_2O and K_2O they are known as foidite. The foidite group is silica-undersaturated and contains the unusual feldspathoid minerals such as nepheline and leucite, rather than the more common feldspar. The distinction between the sodium-rich foidite (nephelinite)—which characterises the EARS—and

the potassium-rich foidite (leucite) cannot be made on this diagram. The even more unusual carbonatite magmas, an extreme type of foidite in which the silicates are almost entirely displaced by carbonates, also cannot be plotted on the TAS diagram. The parental magmas to alkaline rocks are typically generated by melting relatively small volumes of mantle rock, or due to partial melting at greater depths in comparison to the more abundant sub-alkaline magmas which are found in many other volcanic provinces, e.g. Hawaii.

Three broad groups of alkaline magmas are recognised in the Kenyan section of the Gregory Rift (Baker 1987). Each group reveals a discrete fractionation path over time (Fig. 5.8b):

- **Group (i) magmas** of the pre-rift volcanism (22–12 Ma) evolve from nephelinite to phonolite, but can also include carbonatite;
- **Group (ii) magmas** of the half-graben stage (12–4 Ma) evolve from alkali basalt/basanite to phonolite and
- **Group (iii) magmas** of the full graben stage (4 Ma Holocene) evolve from transitional basalt through trachyte to alkali rhyolite.

In a generalised way, the plateau-style volcanics located in the rift valley of central Kenya, as well as on the Western Rift Platform are dominated by Group (i) magmas. The Group (ii) magmas occur in plateau-style lavas both within and adjacent to the rift valley. Volcanics in the rift valley of southern Kenya and on the Eastern Rift Platform are mostly derived from Group (iii) magmas. The discrete edifices and cones in central/southern Kenya, however, reveal no definite compositional trends related to age.

The Older Volcanism of northern Tanzania is dominated by Group (ii) magmas, consistent with the half-graben stage. Examples include most of the Ngorongoro Volcanic complex and the discrete cones of Ketumbeine, Monduli, Oldoinyo Samba, Shira and Tarosero. The Essimingor Volcano and some cones in the Ngorongoro complex, however, are anomalous as they include nephelinite–phonolite [Group (i) magmas]. The Younger Volcanism is dominated by either Group (ii) magmas, e.g. the Burko, Hanang, Gelai and Mawenzi Volcanoes or Group (i) magmas, e.g. the Kerimasi, Mosonik, Kibo and Oldoinyo Lengai Volcanoes. Meru is unusual in that it is comprised of alkali basalt [Group (ii) magmas] and nephelinite [Group (i) magmas]. The reversal towards the Group (i) magmas (nephelinite–

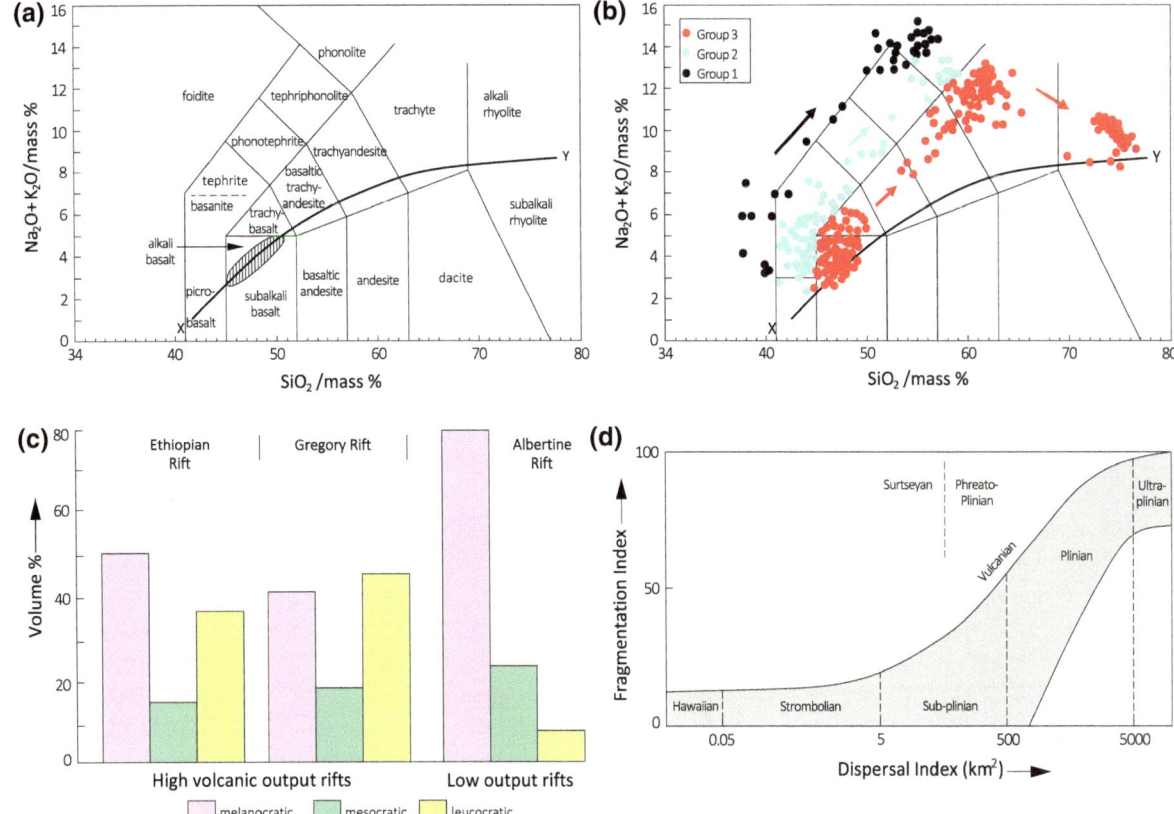

Fig. 5.8 a The TAS diagram, a plot of total alkali (sodium and potassium oxides) versus silica, is a popular method of categorising volcanic rocks. The line X-Y is used to separate the alkaline group (above) from the sub-alkaline group (below). Basaltic rocks include three categories, alkali basalt, transitional basalt (shaded area straddling the division) and subalkali basalt. The unusual carbonatite volcanic rocks cannot be plotted on this diagram, although they can be considered as an extension of the highly alkaline foidite group (which includes nephelinite and leucite); **b** Lavas from the Gregory Rift, Kenya plotted on a TAS diagram define three groups (after Baker 1987): Group (i) nephelinite–phonolite (all samples plotting in the foidite field are nephelinite); Group (ii) alkali basalt/basanite–phonolite; and Group (iii) transitional basalt–trachyte–alkali rhyolite. Arrows illustrate evolution trends with decreasing age; **c** The relative output of the three branches of the EARS expressed by relative proportions of volcanics subdivided by colour index (after Barberi et al. 1982). In a broad sense,

the melanocratic, mesocratic and leucocratic groups correspond with chemical classifications, basic (>52% SiO_2), intermediate (52–63% SiO_2) and acid (>63% SiO_2), respectively; **d** Volcanoes that generate pyroclastic fall deposits can be divided into groups based on a plot of fragmentation index (percentage of deposit with grains finer than 1 mm) against the area of dispersal (after Walker 1973 and Wright et al. 1980). The relatively quiescent Hawaiian and Strombolian styles of eruptions are dominated by lavas and ash-fall deposits with very limited distribution, whereas the catastrophic Plinian and ultra-Plinian eruptions have larger components of pyroclastics that disperse over great distances. Shaded area includes the bulk of eruptions. Volcanoes located above the line are associated with either relatively short-lived explosive eruptions (Vulcanian) or systems that have interacted with water (Surtseyan or Phreatoplinian). Most of the volcanoes associated with the Gregory Rift fall within the sub-Plinian and Plinian divisions phonolite and carbonatite) in the youngest centres has been interpreted as indicating a waning of the EARS in northern Tanzania.

Box 5.1: Calderas and Plinian Eruptions

Many of the volcanoes in East Africa are associated with the explosive styles of volcanism that can produce catastrophic events such as calderas and pyroclastic flows. Whereas, craters are relatively small, typically circular features that occur at the summit of cones or juxtaposed with side vents, calderas (from the Latin calderia or 'boiling pot') are much larger, often basin-shaped features that can dominate the entire volcano. Volcanic craters can form by means of constructive processes where a rim builds up by repeated eruption of lavas and ashes. Calderas, however, generally form where an entire cone has collapsed due to rapid depletion of the underlying magma chamber. Calderas are structurally controlled and may be defined by large faults. Calderas can also form as a consequence of the structural collapse of large sectors of a cone. Calderas are not always recognised as they can extend far beyond the eroded relict of an original cone. Moreover, rejuvenation of volcanic activity can occur within the centre of a caldera, with the formation of new cones.

The observations and letters sent by Pliny the younger to the historian Tacitus describing the 79 AD eruption of Vesuvius, Italy, have been incorporated into geological science. Pliny observed the initial stages of the eruption and sketched the enormous ash column that he compared to the shape of an 'umbrella pine'. His letters also described details of the death of his uncle, Pliny the elder, as resulting from asphyxiation and poisonous gases associated with the subsequent pyroclastic flows. Plinian eruptions typically commence with sustained ash columns and pyroclastic flows. They may be followed by volcano-tectonic collapse and the formation of calderas. They are typically associated with stratovolcanoes such as those found

on the Eastern Platform of the Gregory Rift. One of the most hazardous features of this eruptive style is the occurrence of near-instantaneous basal surges prior to pyroclastic events. Roof collapse due to the enormous volumes of ash erupted is an additional hazard. The 79 AD eruption occurred after 400 years of quiescence. The absence of intermediate events prohibits accurate predictions and the fertile soils, and enhanced rainfall of many volcanoes encourages settlement.

5.10 Volcanic Output and Explosiveness

The Gregory and Ethiopian Rifts are distinguished from the Albertine Rift by their extremely high volcanic output (Fig. 5.8c). The intensity of the volcanism is evident from the extensiveness of plateau-style lava fields, including faulted terranes upon which younger cones are built. High-output rifts such as the Gregory Rift typically reveal a bimodal distribution in which the volume of melanocratic lavas/ashes (basalts) is matched by the abundance of leucocratic lavas/ashes (phonolite, trachyte and rhyolite). Continental rifting generally results in highly explosive, Plinian styles of volcanism (Barberi et al. 1982) and the Gregory Rift is no exception (Fig. 5.8d). Plinian-style eruptions are associated with extensive pyroclastic deposits (Walker 1973; Wright et al. 1980) and giant calderas (Box 5.1). Menengai (Kenya) and Ngorongoro (northern Tanzania) are two of the largest and best-preserved calderas on Earth. The relative explosiveness of the Younger Volcanism in northern Tanzania is exemplified by a high proportion of pyroclastics.

5.11 Other Volcanic Features

Despite crustal extension creating plumbing systems that could facilitate upward migration of magma, volcanism in the Gregory Rift was fed through discrete feeder systems, an unusual feature first recognised by Gregory (1896). Most of the rift-related faults are tightly sealed and could not have acted as conduits. The longevity of feeders is apparent from the occurrence of giant stratovolcanoes. Clusters of feeders produced overlapping volcanic complexes. Le Gall et al. (2008) have suggested that significant differences between the structural regimes within the northern Tanzanian divergence have resulted in two different volcanic styles. Orthogonal faulting near Lake Natron produced small, discrete cones. Oblique rifting with an east–west extension in the centre of the divergence resulted in some of the worlds' largest volcanic edifices.

Many volcanoes in the Gregory Rift experienced periodic sector collapse, typical of highly explosive systems. Sector collapse can produce extensive debris avalanche deposits (DADs), as described, for example, by Van Wyk de Fries and Delcamp (2015). DADs are distinguished from lahars (superficial volcanic deposits mobilised by water) as they result from the partial or total collapse of cones with transport of the resulting debris independent of water. Sector collapses are associated with highly explosive volcanoes and can produce huge horseshoe-shaped calderas of which Mount Meru is one of the most spectacular examples known (Delcamp et al. 2015).

5.12 Sedimentary Basins

Sedimentary basins associated with the Gregory Rift formed in response to persistent down-faulting in and adjacent to the rift valley, as well as from localised warping of the regional plateau (Baker 1958; Dawson 2008). Most basins in Kenya are too small to depict on the regional map, with the principal exception being the Amboseli Basin, which is associated with a large warp on the Eastern Rift Platform (Fig. 5.2). As topography is persistently reshaped by rifting and volcanism, lakes and river channels may be abandoned, or covered by later sequences, e.g. as reported from the Lake

Baringo basin (Renault and Tiercelin 1994). Sediment-filled basins are widespread in the northern Tanzania divergence where they may occur in second-order grabens and small flexures (Fig. 5.5). The sediments at Oldupai Gorge and Laetoli formed in palaeo-lakes created by warping of the Serengeti Plains, to the west of the Eyasi half-graben. Most sedimentary basins associated with the EARS started to form during the Late Miocene. The exposed sediment fill is, however, typically Pleistocene or Holocene in age. Basins are infilled with sequences of poorly consolidated clays, sands, volcanic ashes and volcaniclastics. The Suguta Valley of central Kenya is a Pleistocene-age basin in which sediments accumulated in a large lake, whether this lake persisted into the Holocene is not known. This basin also reveals small basaltic lava flows.

5.13 Rift Valley Lakes

Most lakes in the Gregory Rift are shallow and alkaline, with distinctive ribbon or finger shapes. Examples include Lakes Eyasi, Bogoria, Magadi, Manyara and Natron (Figs. 5.2 and 5.5). They occur in dry, desolate sections of the rift where evaporation rates are high and basins have restricted catchments. Outward tilting of the rift platforms has resulted in most rivers in the region draining away from the rift valley (Sect. 3.3). Lakes Magadi and Natron in southern Kenya and northern Tanzania, respectively, are extreme examples of alkaline lakes located in a particularly dry and desolate area (Baker 1958). Vegetation is sparse, rainfall erratic and temperatures are regularly over 40 °C. The lakes have maximum surface areas of over 100 km^2, but during the dry season extensive salt flats develop as the depth is typically only a few metres. The Lake Magadi brines are extraordinarily toxic with a pH as high as 12 (Jones et al. 1977; Eugster 1980). Lakes Naivasha and Baringo are anomalous as they occur in slightly larger basins, have rounded shapes, are significantly deeper and are mostly freshwater.

Plate 5.2 Exposures in the Engare Sero Gorge near Lake Natron include (**a**) A massive layer of basaltic lava (black) that overlies a bed of nephelinitic volcanic ash (pale green-grey); (**b**) A nephelinite lava with abundant lithic fragments has a characteristic pale green-grey colour

Plate 5.3 Exposures in the Engare Sero Gorge near Lake Natron include (**a**) Basaltic lava with prominent columnar jointing; (**b**) Nephelinitic breccia dyke that cuts a layer of basaltic lava

Plate 5.4 Exposures in the Engare Sero Gorge near Lake Natron include (**a**) Debris avalanche deposits, with considerable variability of block size; (**b**) The angular nature of blocks of varying compositions (lavas, ashes, etc.) is the characteristic of debris avalanche deposits

References

Bailey, D. K. (1974). Continental rifting and alkaline magmatism. In H. Sorensen (Ed.), *The alkaline rocks* (pp. 148–159). New York: Wiley.

Baker, B. H. (1958). Geology of the Magadi area. *Geological Survey of Kenya Report, 42*, 81 p.

Baker, B. H. (1987). Outline of the petrology of the Kenyan Rift alkaline province. In: Fitton, J. G. & Upton, B. G. J. (Eds.), *Alkaline igneous rocks* (Vol. 30, pp. 293–311). Geological Society of London Special Publication.

Baker, B. H., Mohr, P. A., & Williams, L. A. J. (1972). Geology of the Eastern Rift system of Africa. *Geological Society of America Special Paper, 136*, 67 p.

Baker, B. H., & Wohlenberg, J. (1971). Structure and evolution of the Kenyan Rift Valley. *Nature, 229*, 538–542.

Barberi, F., Santacroe, R., & Varet, J. (1982). Chemical aspects of rift magmatism. In: Palmason, G. (Ed.), *Continental and oceanic rifts: final report of inter-union commission on geodynamics workshop group 4, "Continental and Oceanic Rifts"* (Geodynamics Series 8, pp. 223–258). Washington: American Geophysical Union.

Chorowicz, J. (2005). The East African Rift system. *Journal of African Earth Sciences, 43*, 379–410.

Dawson, J. B. (2008). The Gregory Rift Valley and Neogene-recent volcanoes of Northern Tanzania. *Geological Society London Memoir, 33*, 102 p.

Delcamp, A., Delvaux, D., Kwelwa, S., Macheyeki, A., & Kervyn, M. (2015). Sector collapse events at volcanoes in the North Tanzanian divergence zone and their implications for regional tectonics. *Geological Society of America Bulletin, 128*, 169–186.

Ebinger, C., Poudjom-Djomani, Y., Mbede, E., Foster, A., & Dawson, J. B. (1997). Rifting Archaean lithosphere: The Eyasi-Natron rifts, East Africa. *Journal of the Geological Society of London, 154*, 947–960.

Eugster, H. P. (1980). Lake Magadi, Kenya and its Pleistocene precursors. In A. Nissenbaum (Ed.), *Hypersaline brines and evaporitic environments* (pp. 195–232). Amsterdam: Elsevier.

Evans, A. L., Fairhead, J. D., & Mitchell, J. G. (1971). Potassium-argon ages from the volcanic province of Northern Tanzania. (Vol. 229, pp. 19–20) London: Nature Physical Science.

Gregory, J. W. (1896). *The Great Rift Valley* (421 p). London: John Murray.

Jones, B. F., Eugster, H. P., & Rettig, S. L. (1977). Hydrochemistry of the Lake Magadi basin, Kenya. *Geochimica et Cosmochimica Acta, 41*, 53–72.

Le Gall, B., Nonnotte, P., Rolet, J., Benoit, M., Guillou, H., Mousseau-Nonotte, M., et al. (2008). Rift propagation at craton margin: distribution of faulting and volcanism in the north Tanzanian divergence (East Africa) during Neogene times. *Tectonophysics, 448*, 1–19.

Mana, S., Furman, T., Carr, M. J., Mollel, G. F., Mortlock, R. A., Feigenson, M. D., et al. (2012). Geochronology and geochemistry of the Essimingor Volcano: melting of metasomatized lithospheric mantle beneath the northern Tanzania divergence zone (East African Rift). *Lithos, 155*, 310–325.

Renaut, R. W., & Tiercelin, J.-J. (1994). Lake Bogoria, Kenya Rift Valley: a sedimentological overview. In: Renaut R. W. & Last W. M (Eds.), *Sedimentology and geochemistry of modern and ancient Saline Lakes* (Vol. 50, pp. 101–123). SEPM Special Publication.

Rogers, N., MacDonald, R., Fitton, J. G., George, R., Smith, M., & Barreiro, B. (2000). Two mantle plumes beneath the East African Rift System: Sr, Nd, and Pb isotope evidence from Kenya Rift basins. *Earth and Planetary Science Letters, 176*(3–4), 387–400.

Walker, G. P. L. (1973). Explosive volcanic eruptions—a new classification scheme. *Geologische Rundschau, 62*, 431–446.

Woolley, A. (2001). *Alkaline rocks and carbonatites of the world. Part 3: Africa* (372 p). Geological Society of London.

Wright, J. V., Smith, A. L., & Self, S. (1980). A working terminology of pyroclastic deposits. *Journal of Volcanology and Geothermal Research, 8*, 315–336.

Wyk, Van, de Vries, B., & Delcamp, A. (2015). Volcanic debris avalanches. In T. Davies & J. F. Shroder Jr. (Eds.), *Landslides hazards, Risks and Disasters* (pp. 131–157). San Francisco: California Academic Press.

Late Pleistocene Ice Ages and the Holocene Epoch

Abstract

The Late Pleistocene Ice Ages were a worldwide series of events that resulted in extreme climatic cycles. In the vicinity of the Gregory Rift, climate changes were accompanied by intensive volcanism. In comparison, the rapid explosion in population numbers of *Homo sapiens* that ushered in the Holocene occurred in a time of modest climatic fluctuations and less intense volcanism. Significant climatic cycles did, however, occur in the Holocene of East Africa. Evidence includes ice cores from Kibo (Kilimanjaro) and changes in the size and depth of lakes in the rift valley. The Early Holocene was dominated by the African Humid Period, a relatively hot and wet phase that lasted from 11,700 BP until approximately 6,000–5,000 BP. This period was interrupted by an intensely dry phase at approximately 8,300 BP. The Late Holocene included a second extremely dry phase (4,000–3,700 BP) which affected civilisation in northern Africa and the Middle East so severely that it is known as the First Dark Age. During the hotter and more humid periods of the Early Holocene, giant palaeo-lakes developed in the Gregory Rift Valley. Conversely, many lakes dried up during the intervening arid phases. Since 3,700 years, the climate established similar patterns to those observed today although minor, relatively short-lived cycles are recognised, including the Medieval Warming and the Little Ice Age. Volcanism in the Holocene, which can be categorised as active or dormant, occurred both in the Gregory Rift Valley and on the Eastern Rift Platform. Several catastrophic events are recognised which would have affected evolution of the hominins as well as growth of some of the montane forests.

Keywords

Active volcanism • Climatic cycles • Holocene
Montane forest • Palaeo-lake • Pleistocene Ice Ages

Photographs not otherwise referenced are by the author.

© Springer International Publishing AG, part of Springer Nature 2018
R. N. Scoon, *Geology of National Parks of Central/Southern Kenya and Northern Tanzania*, https://doi.org/10.1007/978-3-319-73785-0_6

Plate 6.1 The slopes of Mount Meru are covered by extensive montane forests which regenerated after the catastrophic eruption of approximately 7,800–7,000 BP

6.1   Introduction

The Late Pleistocene was marked by periods of extreme climatic fluctuations, global phenomena known as the Ice Ages. Excessively cold and dry periods alternated with warmer and wetter periods. Large sections of the Gregory Rift in central/southern Kenya and northern Tanzania were subjected to continued rifting and intense volcanism during this period. The Holocene ushered in an epoch of less intense climatic changes and relatively modest seismic activity and volcanism. Ice cores drilled into the Northern Ice field of Kibo (Kilimanjaro) provide important evidence with regard to climatic cycles in the Holocene of this area. These climatic cycles had a considerable effect on the size and depth of palaeo-lakes within the Gregory Rift. Volcanism, despite being less pervasive in comparison to the Plio-Pleistocene, includes a dozen or so active volcanoes. The caldera events associated with the Menengai Volcano and Mount Meru were catastrophic events which would have influenced the local climate. Some of the montane forests that occur on the volcanic cones may, therefore, be younger than appreciated (Plate 6.1).

6.2   Late Pleistocene Ice Ages

The Ice Ages of the Late Pleistocene included at least seven relatively long glacial epochs separated by notably shorter warmer periods. The half-dozen or so early glaciations occurred between approximately 0.68 and 0.13 Ma. The Main Ice Age occurred between 0.11 Ma and 12,000 BP, lasting for approximately 100,000 years and including several glacial peaks. The most recent of the glacial peaks, the Last Glacial Maximum, occurred at 20,000 BP. The effects of the Ice Ages were particularly extreme in the northern latitudes, with large areas of Europe and North America being covered by ice caps and glaciers, but this was a global phenomenon which influenced the climate of even the tropical regimes. Changes in temperature and precipitation associated with the Ice Ages may be ascribed to the precession cycle of some 23,000 years (Precession is a regular change in the orientation of the axis of a rotating sphere, such as Earth).

The Late Pleistocene of East Africa included periods of intense cold and dry ('ice ages'), that alternated with warmer and wetter periods (Osmaston 2004), but the effects were not as extreme as in northern latitudes. Climatic cycles in East Africa were tempered by the equatorial setting and ice developed only as icecaps and slope glaciers on the larger peaks, e.g. Kilimanjaro (5,895 m), Mount Kenya (5,199 m), Mount Meru (4,562 m) and Mount Elgon (4,321 m).

6.3 Holocene Climatic Cycles

The distribution of ice on the East African Mountains has changed repeatedly since the Main Ice Age (Thompson et al. 2002; Osmaston 2004; Hardy 2011; Cullen et al. 2013; Pepin et al. 2014). An extraordinarily dry period at around 12,000 BP is thought to have caused most icecaps to entirely disappear. Smaller icecaps reformed 300 years later during the start of the Holocene epoch at 11,700 BP. The Early Holocene was dominated by a warmer and wetter climate, the African Humid Period. This lasted until approximately 6,000–5,000 BP but was interrupted by an intensely dry period at approximately 8,300 BP which coincided with rapid cooling of the North Atlantic Ocean. An increase in the sodium and fluorine in aerosols trapped in the ice of the Northern Ice field of Kibo provides corroborating evidence of this and other dry periods during the Holocene. Sodium and fluorine are important constituents of alkaline lakes which pollute the atmosphere when lakes dry up and the salt crusts disintegrate.

The most important of the dry periods recognised in East Africa occurred during the Late Holocene at approximately 4,000–3,700 BP. This resulted in the rapid recession of the mountain ice caps and the drying of lakes. This period is known to historians as the First Dark Age as the effects were widespread and affected civilisations around the Nile River and in the Middle East. After approximately 3,700 BP, the climate settled into a similar pattern to that currently observed (including the biannual monsoon). Within this period, however, 300–400 year-long climatic cycles, which are probably universally applicable, also occur. The two most well known of these relatively short cycles are the Medieval Warming (800–1250 AD) and the Little Ice Age (1250–1850 AD).

6.4 Monsoon

The climate of East Africa is currently dominated by the twice-yearly passage of the inter-tropical convergence zone or Indian Ocean monsoon. Two rainfall seasons (March–May and October–November) are generally experienced. Projections on global warming suggest an increase in precipitation in East Africa and there is general consensus that the rainy seasons have in recent years been less well defined. Decade-long trends that overprint the 300–400 year cycles may reflect sensitivity to subtle variations in, for example, cross-equatorial heat transport by ocean currents.

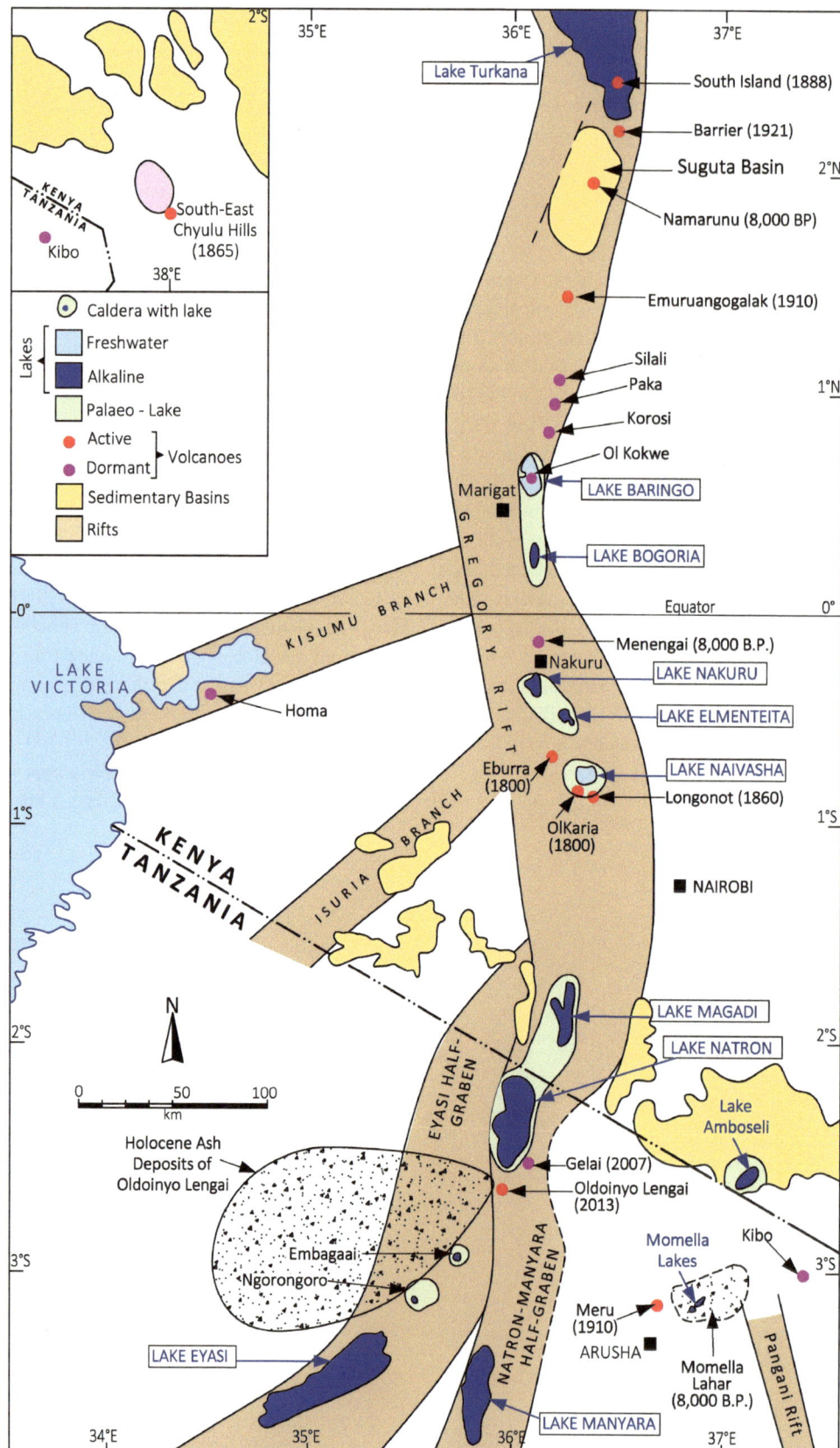

Fig. 6.1 Map showing some features of the Holocene geology, including palaeo-lakes and active and dormant volcanoes, in the Gregory Rift Valley of central/southern Kenya and northern Tanzania. The inset shows the location of the Chyulu Hills in southern Kenya

6.5 Palaeo-Lakes

Many of the ribbon-shaped, alkaline lakes of the rift valley experienced higher water levels during the humid climate of the Early Holocene. The enlarged, palaeo-lakes were far deeper (>100 m) than current levels and were dominated by freshwater (Fig. 6.1). Fossil evidence from lacustrine and volcaniclastic deposits have shown that Lakes Magadi and Natron were connected during the Early Holocene (and parts of the Late Pleistocene). The Magadi–Natron palaeo-lake was freshwater and contained species of fish that are extinct. Connections also occurred between Lakes Baringo–Bogoria and Nakuru–Elmenteita–Naivasha. Lake Naivasha was possibly three times larger than it is today and considerably deeper (150 m compared with a current average of 6 m). During the arid periods described above, i.e. at 8,300 BP and 4,000–3,700 BP, palaeo-lakes shrunk and became increasingly alkaline. Some palaeo-lakes even dried up completely.

6.6 Historical Lakes

Historical records that extend to the beginning of the previous century reveal that the rift valley lakes are extremely sensitive to even the smallest climatic fluctuations. Considerable variability is documented in both extent and depth. Lake Naivasha, for example, was twice as large in the 1920s as in the 1960s. Many lakes reported sharp rises in the 1960s but there has been a gradual decline since then, probably due to a natural cycle. Higher water levels since 2013, due to increased rainfall, have caused localised flooding of the foreshore around some lakes.

6.7 Active Volcanism

Most Holocene volcanoes of the Gregory Rift are categorised as active (Global Volcanism Programme 2016) or dormant, although this distinction is vague and has little scientific credibility (Table 6.1). From north to south, the following volcanoes are identified in central/southern Kenya (Fig. 6.1). The active South Island Volcano with its spectacular crater (Plate 6.2a) last erupted in 1888. The active Barrier Volcano last erupted in 1921, the dormant Namarunu Volcano in 8,000 BP and the active Emuruangogolak Volcano in 1910. The main caldera event of the giant Menengai Volcano occurred at 8,000 BP, with the most recent, relatively small-scale activity occurring approximately 200 years ago. A geothermal field at Menengai has been linked to an active magma chamber at relatively shallow depth. The Eburru and Olkaria complexes near Lake Naivasha both experienced eruptions approximately 200 years ago. The active Mount Longonot Volcano is estimated to have last erupted in 1860. Some of the volcanic centres in the Chyulu Hills (inset, Fig. 6.1) are part of an active system. The Shaitani and Chaimu flows and cones are estimated to have erupted in 1865–1866.

The most active volcano in northern Tanzania is Oldoinyo Lengai, which typically erupts every 15–20 years. The last significant eruption here was in 2013. The adjacent Gelai Volcano showed minor signs of recent activity in 2007. The Kibo component of Kilimanjaro is generally considered as a dormant volcano but there is no evidence to suggest it is not part of an active system: the Reusch Crater on the summit plateau reveals evidence of geothermal activity. The most recent activity at the Meru Volcano occurred in 1910, although the caldera event occurred at approximately

Table 6.1 Table of active and dormant volcanoes

Volcano/complex	Location	Latitude (north/south)	Longitude (east)	Height (m)	Last eruption	Active/dormant
Central/Southern Kenya						
South Island	Rift valley	2.36N	36.36	800	1888	**Active**
Marsabit	Eastern Platform	2.15N	37.57	1707	Holocene?	Dormant
Barrier	Rift valley	2.20N	36.37	1032	1921	**Active**
Namarunu	Rift valley	1.58N	36.26	817	8550 BP	Dormant
Emuruangogolak	Rift valley	1.29N	36.19	1328	1910	**Active**
Silali	Rift valley	1.10N	36.12	1528	7050 BP	Dormant
Paka	Rift valley	0.55N	36.12	1697	8050 BP	Dormant
Korosi	Rift valley	0.46N	36.07	1446	Holocene?	Dormant
Ol Kokwe[a]	Rift valley	0.36N	36.04	980	Holocene?	Dormant
Nyambeni Hills	Eastern platform	0.23N	37.87	750	Holocene?	Dormant
Menengai	Rift valley	0.12S	36.04	2278	8050 BP	Dormant
Homa	Kisumu Rift	0.23S	34.30	1751	Holocene?	Dormant
Eburru	Rift valley	0.39S	36.12	2856	(1800)[b]	**Active**
Olkaria	Rift valley	0.53S	36.18	2434	(1800)[b]	**Active**
Longonot	Rift valley	0.55S	36.27	2776	1860	**Active**
SE Chyulu Hills	Eastern Platform	2.50S	38.00	2188	1865	**Active**
Northern Tanzania (NT)						
Gelai	Rift valley	2.38S	36.06	2942	(2007)[c]	Dormant
Oldoinyo Lengai	Rift valley	2.46S	35.54	2890	2013	**Active**
Kibo	NT divergence	3.04S	37.21	5895	Holocene?	Dormant
Mount Meru	NT divergence	3.15S	36.45	4565	1910	**Active**

[a]Part of Korosi Volcano
[b]Estimation
[c]Fissure eruption that didn't reach the surface
Some Holocene Volcanoes (Holocene?) have no accurate dates of most recent eruptions

7,800–7,000 BP. Catastrophic volcanism may trigger climatic cycles (by pollution of the atmosphere, for example), and the Plinian-style eruptions of the Menengai and Meru Volcanoes during the Early Holocene would have severely impacted on any local human populations.

6.8 Montane Forests

The extensive montane forests on many of the volcanic cones in central/southern Kenya and northern Tanzania are associated with localised areas of higher rainfall in comparison to the relatively dry plateaus and valleys. Whether these are remnants of forests that formerly covered large parts of East Africa, or are younger, more localised features has not been resolved. Some forests must be relatively young due to the recent age of some of the catastrophic volcanism, e.g. Mount Meru. Typically, the lower and central slopes of mountains are girdled by forests of giant podocarpus trees together with wild olive and juniper. The lowermost parts of the forests merge into savannah woodlands or grassy plateaus, with the upper parts giving way to belts of bamboo and heathland that may include endemic giant groundsels and lobelias. Some forests include belts of cycads reminiscent of a Jurassic-age environment. The forests and upper heathlands include extensive flowers (including endemic orchids) and constitute some of the most important botanical localities known. The forests of Mount Elgon are particularly well known in this regard. Some plant species occurring high on the East African Mountains are only found in one other locality, the Cape Floral Kingdom of South Africa. Whether this is evidence of connectivity between these areas prior to the development of relatively young deserts is speculative. Mountain ecosystems are unusually sensitive to natural and anthropogenic land-use changes so that replacement by agriculture and settlements leads to erosion, disruption of water sources and drying up of rivers (Whiteman 2000). Moreover, forests are globally important as they assist with modifying greenhouse gases, water supply and species diversity.

Mountains enable rapid detection of climatic variations as changes in vegetation and hydrology occur over short distances. The highlands of East Africa were probably resilient to exploitation until recent times, but land degradation is now threatening both park ecosystems and farming communities.

Plate 6.2 **a** The active Nabiyotum Crater, Lake Turkana, is a spectacular site. Photograph by Christian Strebel; **b** The summit crater of Oldoinyo Lengai prior to the 2008 eruption looking south towards the extinct Kerimasi Volcano. *Source* Public domain internet site https://uploads.disquscdn.com/images/ca.jpg

References

Cullen, N. J., Sirguey, P., Mölg, T., Kaser, G., Winkler, M., & Fitzsimons, S. J. (2013). A century of ice retreat on Kilimanjaro: The mapping reloaded. *The Cryosphere, 7,* 419–431.

Global Volcanism Program (2016). *Smithsonian Institute, Washington.* http://volcano.si.edu/index.cfm/gup_about.cfm.

Hardy, D. R. (2011). Kilimanjaro. In V. P. Singh, P. Singh, & U. M. Haritashya (Eds.), *Encyclopedia of snow, ice and glaciers, Encyclopedia of earth sciences series* (pp. 672–679). Berlin: Springer.

Osmaston, H. (2004). Quaternary glaciations in the East Africa Mountains. In J. Ehlers & P. I. Gibbard (Eds.), *Developments in quaternary sciences, quaternary glaciations extent and chronology part III: South America, Asia, Africa. Australasia, Antarctica* (Vol. 2C, pp. 139–150). Amsterdam: Elsevier.

Pepin, N. C., Duane, W. J., Schaefert, M., Pike, G., & Hardy, D. (2014). Measuring and remodelling the retreat of the summit icefields on Kilimanjaro, East Africa. *Arct Antarct Alp Res, 46,* 905–917.

Thompson, L. G., Mosley-Thompson, E., Davis, M. E., Henderson, K. A., Brecher, H. H., Zagorodnov, V. S., et al. (2002). Kilimanjaro ice core records: Evidence of Holocene climate change in tropical Africa. *Science, 298,* 589–593.

Vye-Brown, C., Crummy, J., Smith, K., Mruma, A., & Kabelwa, H. (2014). Volcanic hazards in Tanzania. In: *British Geological Survey Open File Report OR/14/005,* 29 p.

Whiteman, C. D. (2000). *Mountain Meteorology: Fundamentals and applications* (355 p). New York: Oxford University Press.

Serengeti National Park

Abstract

The Serengeti National Park is famous for the open plains (savannahs) that characterise many of the regional plateaus in East Africa. The Serengeti Plains reveal a pronounced westward slope, with elevations decreasing from approximately 1,850 m near the Ngorongoro Highlands to less than 1,000 m near Lake Victoria. The higher elevation in the vicinity of the Ngorongoro Highlands may be due to uplift on the margins of the rift valley. Large parts of the Serengeti are underlain by some of the oldest rocks on Earth, namely greenstones and granite-gneiss of the Archaean-age Central Africa craton. These ancient crystalline rocks crop out in small koppies and elongate hills. The greenstones give rise to black clay soils, almost impossible to drive over, and the granite-gneiss to light-coloured sandy soils. Outcrops of the Mozambique Belt can also be observed as can sedimentary rocks of the Ikorongo Group, the latter occurring as linear ridges in central parts of the park. The age of the Ikorongo is uncertain and they may straddle the boundary between the Neoproterozoic and Cambrian. The youngest rocks occur as a thin cover of volcanic ash; they are restricted to the eastern plains. Three ecosystems are generally identified on the Serengeti Plains, the wooded savannahs and long grass plains of the western and northern areas, and the short grass of the eastern plains. The migration of more than two million grazers is largely controlled by rainfall patterns, but the huge herds that concentrate for a key period in the relatively arid eastern plains are reacting to recent geological events. The nutrient-rich soils that support the short grasses are derived from ashes erupted from the Oldoinyo Lengai Volcano. The Serengeti Plains are drained by large, westward-flowing rivers which are associated with sediment-filled, sinuous channels. Some of the rivers, e.g. the Mara have carved substantial channels down to the bedrock and the sight of groups of several hundred wildebeest crossing the rocky riverbeds during the migration is an unforgettable sight.

Keywords

Archaean • Central African craton • Granite-gneiss Greenstones • Migration • Regional plateau Volcanic ash

---

Photographs not otherwise referenced are by the author.

Plate 7.1 Koppie or small hill of Archaean-age granite-gneiss, Eastern Serengeti Plains

7.1 Introduction

Large parts of the regional plateau that constitutes the world-famous Serengeti National Park have been protected since 1929. One of the most distinctive physiographic features is the presence of small hills, or koppies, most of which are inselbergs of ancient resistant rocks that project from the plateau (Plate 7.1). The Serengeti is justifiably famous for one of the most fascinating natural spectacles on Earth, the migration of more than two million large grazers. The park covers an area of some 14,763 km^2 with many of the surrounding areas also protected in additional reserves (Fig. 7.1). The dominant feature of this large and relatively undisturbed ecosystem is the seemingly endless Serengeti Plains, part of the regional plateau that extends for over 100 km between the Ngorongoro Highlands and Lake Victoria. The plateau has an imperceptible westward slope, with the elevation decreasing from 1,850 m in the east to 925 m in the west. The elevation change is superimposed on the regional plateau by the EARS: the higher elevation in the east is related to uplift adjacent to the Eyasi half-graben and the lower elevation in the west to the warp associated with the Lake Victoria Basin (Sect. 3.8).

The Serengeti Plains are characterised by savannah grasslands and thorn bush that can be subdivided into three main ecosystems (Inset, Fig. 7.1). The long grass plains with scattered acacia trees dominate the central area. Areas of thorn bush, with minor gallery forests occurring along river channels, and clay pans occur in the northern part and western corridor. The almost treeless short grass plains are restricted to the eastern plains.

7.2 Approaches

The Serengeti is generally approached by a road that winds down from the Ngorongoro Highlands onto the dry, eastern plains in the vicinity of the Naabi Gate (Fig. 7.1). Despite crossing a major rift-related fault downthrown to the east (Fig. 5.4), the Serengeti Plains occur at a *lower* elevation than the Ngorongoro Highlands. The reason for this unusual situation is the great thickness of volcanic rocks that were built up within the Eyasi half-graben from Pliocene-age volcanism. Alternative approaches are to enter the north-eastern Serengeti near the Lobo Lodge, a route that crosses the rugged Loliondo Wilderness Area near the Kenyan border (this road connects with the rift near Lake

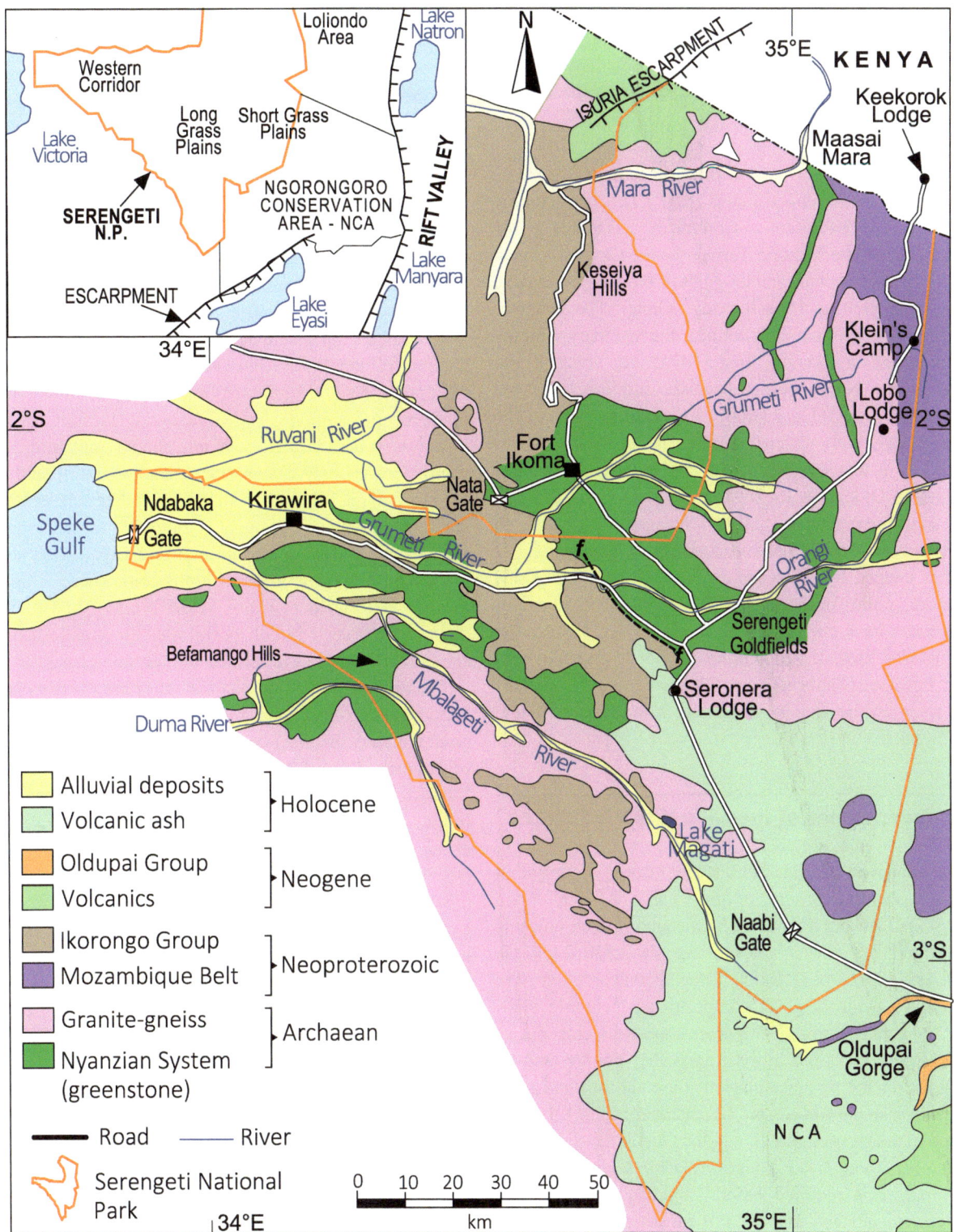

Fig. 7.1 Geological map of the Serengeti National Park simplified and updated from various sources including Pickering (1964), MacFarlane (1965) and Barth (1990). The Holocene age of the calcareous ash is after Dawson (2008) and identification of the Ikorongo Group is after Kasanzu and Manya (2010)

Natron), or from Mawenzi via the western corridor and the Ndabaka or Nata Gates.

7.3   Geological Framework

A simplified geological map depicts the principal terranes that make up the Serengeti Plains (Fig. 7.1). The relatively well-exposed Basement Complexes include the Archaean-age Lake Victoria Terrane of the Central African craton (greenstones and granite-gneiss), the Neoproterozoic Mozambique Belt, and the Ikorongo Group. The younger, Neogene-age sequences crop out near the boundary with the Ngorongoro Conservation Area (NCA) i.e. volcanic ash related to the Ngorongoro volcanism and the lacustrine sediments of the Oldupai Group. In addition, parts of the eastern plains, both within the park and in the NCA, are covered by a thin coating of ash derived from Holocene eruptions of Oldoinyo Lengai. The large rivers on the Serengeti Plains, e.g. Grometi, Mara, Mbalageti and Orangi, all flow westward towards Lake Victoria. They display sinuous channels capped by Holocene age alluvial deposits. Thick sequences of alluvium also occur adjacent to the Speke Gulf where rivers have cut down onto the low-lying plains adjacent to Lake Victoria. The Oldupai River is anomalous in that it flows eastwards into the Olbalbal Swamps, and the small rivers in the vicinity of Laetoli flow southwards into Lake Eyasi.

7.4   Greenstones of the Lake Victoria Terrane

Several northwest trending belts of greenstones occur within the Lake Victoria Terrane of the Central African craton (Sect. 4.2). They are dominated by metamorphosed volcanic and sedimentary rocks with an age of 2.810–2.630 Ga (Barth 1990; Dirks et al. 2015). They also include intrusive rocks with a wide range of compositions, including ultramafic and mafic. Some of the greenstone belts crop out as bush-covered ridges but in most areas they are covered by thick soils, including black clays that are almost impossible to drive over in the wet season. Greenstones can be observed in some of the river crossings, e.g. the Orangi River between the Seronera and Lobo Lodges (Plate 7.2a) although an armed ranger is required before alighting from vehicles (bridges and culverts are often used as refuges by large predators). Many of the greenstones reveal a distinctive dark, greenish coloration and a schistose texture. This is diagnostic of basaltic lavas, amphibolite and chlorite schists that have been severely deformed.

7.5   Granite-Gneiss Plutons of the Lake Victoria Terrane

The greenstone belts within the Lake Victoria Terrane are intruded and enclosed by giant, coalesced plutons of granite-gneiss that have an age of 2.720–2.560 Ga (Barth 1990; Dirks et al. 2015). Several intrusive events and multiple phases of deformation in which plutons were partially melted and compressed are recognised (Manya et al. 2006). Minor intrusions of other plutonic rocks, together with segments of schist, quartzite and granulite, occur within the granite-gneiss complex. The granite-gneiss can be observed in the Lobo area where whale-backed outcrops are a distinctive feature (Plate 7.2b). These rocks also form clusters of koppies, e.g. the Moru and Simba Koppies in the vicinity of Seronera. Koppies are a favoured location of large predators (Plate 1.2a). Some guidebooks and information boards erroneously suggest koppies are associated with granitic magma that bulged upwards onto the surface but this is incorrect; they are erosional features that formed when the deep-seated plutons were uplifted and exposed to surface processes. The main phase of uplift commenced at approximately 180 Ma in response to the breakup of Gondwana (Sect. 4.6), several billion years after the granite had both crystallised and been deformed into granite-gneiss.

Weathering of granitic plutons—they are usually massive features—typically creates rounded or exfoliated shapes with giant boulders that may be perched precariously on the crest of koppies (Plate 4.3a). Hard, crystalline rocks with little foliation or bedding are subjected to onion-skin weathering (removal of thin surface layers) when exposed to a marked differential in diurnal temperature. Some granitic bodies in the Serengeti, however, are sufficiently jointed as to form tabular or finger-shaped bodies, also known as tors from type localities in southwest England (Plate 4.3b). The granite-gneiss of the Lake Victoria Terrane is typically coarsely crystalline, with a grey or pinkish colour. Large tabular crystals of feldspar define a porphyritic texture in many localities. The gneissic or banded texture, a consequence of partial melting and deformation of the primary granite, is observed as alternating bands of dark-coloured minerals (mica and hornblende) with light-coloured quartz and feldspar. Criss-crossing veins of granitic pegmatite are an additional feature (Plate 4.2b), as is the presence of xenoliths (from centimetre sized to several metres) of dark-coloured schist. In the open plains the granite-gneiss is covered by poor, light-coloured sandy soils, readily identified from the black clay soils associated with the greenstones.

7.6 Mozambique Belt

Some inselbergs on the Eastern Serengeti Plains are composed of quartzite or granite of the Neoproterozoic age Mozambique Belt. The quartzite crops out in the upper parts of the Oldupai Gorge and can be inspected where the main Serengeti road crosses the river on a low-level bridge (Plate 7.3a). The koppie at Naabi Gate is part of a small granitic pluton. These rocks have not been dated and may be as young as 500 Ma, i.e. Cambrian rather than Neoproterozoic. The Mozambique-age granite can be distinguished from the Archaean granite-gneiss by the presence of a coarse-crystalline fabric and the absence of gneissic textures.

7.7 Ikorongo Group

The Ikorongo Group is a group of sedimentary rocks restricted to northern Tanzania and southwestern Kenya (Fig. 5.2). They formed in fault-bounded basins constrained within the Central African craton (Kasanzu and Manya 2010). Their age is uncertain and they may bridge the gap between the Neoproterozoic and Cambrian. The Ikorongo is dominated by sedimentary rocks, including sandstone, shale and siltstone. The presence of mud-cracks and ripple marks is indicative of deposition in a shallow water environment. Sediments were derived from the juxtaposed Archaean rocks during a regional phase of erosion. The Ikorongo has been subjected to only modest levels of metamorphism and deformation. This unusual feature can be explained by the intracratonic setting in comparison to, for example, the Mozambique Belt which has been intensively metamorphosed and deformed as it is located on the perimeter of the craton. The Ikorongo rocks crop out as low, flat-topped hills, indicative of near-horizontal strata, as seen in the central part of the national park in, for example, the vicinity of Seronera (Plate 7.3b). The ridge on which the Sopa Lodge is constructed consists of rocks of the Ikoronga Group. Another example is the Befamango Hills. The linearity of some ridges is a function of faults that probably define the primary basins, e.g. between the Seronera Lodge and the Orangi River.

7.8 Neogene Age Sequences

The volcanic lavas associated with the Ngorongoro volcanism are restricted to the extreme eastern part of the Serengeti Plains and are described in Chap. 10. The Oldupai and Laetoli Groups, which only crop out in restricted parts of river channels (or gorges) are described in Chap. 11.

7.9 Holocene Volcanic Ash

The Holocene age volcanic ash on the Eastern Serengeti Plains was previously thought to be calcrete, but has been re-interpreted by Dawson (1964; 2008) as semi-consolidated, calcareous ash derived from the Oldoinyo Lengai Volcano. The distribution of this deposit correlates approximately with the short grass plains. The ash has been dispersed over large areas of the NCA and Serengeti Plains by the prevailing easterly winds (Chap. 10). The ash covers the basement complexes and Neogene age deposits as a thin blanket. The thinness of this blanket is evident as even very minor topographic changes cause the underlying (older) rocks to be exposed (in koppies and gorges). A shallow layer of hard pan is located between the calcareous ash and underlying rocks. This results in the water table on the short grass plains being far shallower than on the long grass plains or areas of wooded savannah.

7.10 Rivers and Lakes

The Serengeti is cut by a number of west or northwest flowing rivers such as the Grometi, Mara, Mbalageti and Orangi (Fig. 7.1). They all flow away from the rift valley towards Lake Victoria and can be envisaged as headwaters of the Victoria Nile. Rivers have carved substantial channels in the plateau, some of which have cut down to bedrock, but others reveal alluvium or sand-filled valleys (Plate 7.4a). The park also includes several seasonal lakes, notably Lake Ndutu and Lake Magadi, and minor areas of marshland and swamp.

7.11 Economic Ore Deposits

The Lake Victoria Terrane contains significant mineralisation and some of the greenstones are prospective for gold, nickel and chromium (Grey and Thomas 1964). A number of producing gold mines occur in the Mara Goldfields near Lake Victoria, with several old prospects occurring within the national park. The initial discoveries in the 1930s prompted the 'Mara gold rush' which was described by the well known Kenyan writer Elspeth Huxley as a 'gentlemanly pursuit, more of a hobby than a serious venture' in comparison to earlier, Victorian-age scrambles. The Archaean terranes of northern Tanzania also contain numerous kimberlitic pipes, some of which are diamond-bearing (not shown on the map). The kimberlitic pipes, despite being restricted to the Central African craton are, however, of Cretaceous-age (linked to the breakup of Gondwana). There are currently producing mines near Shinyanga, south of the national park, and there is a prominent cluster of kimberlite pipes near Lake Victoria.

7.12 Migration

The alternation of the wet and dry seasons has a pronounced effect on the ecology of the Serengeti Plains. The onset of the biannual rainy seasons (April–May and September–November) triggers remarkable changes and arid landscapes convert rapidly into lush pastures. The Serengeti migration is currently estimated to include more than 1.5 million wildebeest or white-bearded gnu (*Connochaetes taurinus*), 0.5 million Burchell's zebra, 0.2 million Thomson's gazelle and 18,000 eland (*Taurotragus oryx*). This is one of the great natural spectacles of the world. Migration patterns are not as well established as may be envisaged and the larger grazers are more-or-less continuously moving in vast numbers. Witnessing the herds crossing the major river channels, such as the Mara is an unforgettable site (Plate 7.4b). Typically the herds cross northwards into the Maasai Mara Reserve (Kenya) during July–August and return in September–November. In recent times, the migration patterns have become more erratic, possibly due to climate change. In the early part of the twentieth century, the eastwards migration, instead of stopping in the Ngorongoro region continued as far as the Tarangiri National Park (Fig. 1.1). Calving, the most static time for the wildebeest, usually occurs during January–March on the arid eastern plains. This is supported by the growth of nutrient-rich short grasses. Growth persists for a considerable period due to the shallowness of the water table. Recent geological events have, therefore, influenced the migration, notably the Holocene eruptions of the Oldoinyo Lengai Volcano, and the patterns may not be as ancient as is generally believed. The interrelationship between geology and wildlife in East Africa suggests that the latter react rapidly to geological change (evolutionary or behaviour).

7.13 Koppies

The koppies of granite-gneiss and quartzite described earlier are associated with discrete ecosystems. They include plants and host wildlife not usually seen on the open plains. A notable example is the Klipspringer (*Oreotragus oreotragus* or rock jumper when translated from Afrikaans). One reason for this is that the hardpan on plains adjacent to koppies is less resistant (due to differential weathering) and allows tree roots, for example, to develop more readily. Moreover, the rocky outcrops capture rainwater that in other areas seeps through and is not so readily available for the wildlife. The Serengeti Plains are one of the traditional areas of the Maasai people and cultural imprints include the Gong Rocks (Plate 7.5a) and rock paintings at the Moru Koppies (Plate 7.5b).

Plate 7.2 **a** Archaean-age greenstones are exposed in the bed of the Orangi River where the dark colours and steep dips are characteristic, Serengeti; **b** Archaean-age granite-gneiss forms low, whale-backed outcrops in the Lobo area, Serengeti

Plate 7.3 **a** Quartzite of the Mozambique Belt crops out in the Oldupai River; **b** Ikorongo Group sedimentary rocks are associated with prominent, flat-topped ridges in the Seronera area, Serengeti

Plate 7.4 a Major rivers in the Serengeti, such as the Mara, near the border with Kenya, have carved shallow, alluvium-filled channels in the regional plateau; **b** Wildebeest crossing southwards over the Mara River next to outcrops of Archaean granite-gneiss

Plate 7.5 **a** Carvings into the Archaean-age granite-gneiss at Gong Rock, Moru Koppies, were made by the Maasai for ceremonial purposes, Serengeti; **b** Rock paintings made by the Maasai on Archaean-age granite-gneiss at Moru Koppies, Serengeti

References

Barth, H. (1990). Explanatory notes on the 1:500,000 Provisional geological map of the Lake Victoria Goldfields, Tanzania. *Geologisches Jahrbuch Reihe B, Heft 72,* 59.

Dirks, P. H. G. M., Blenkinsop, T. G., & Jelsma, H. A. (2015). The geological evolution of Africa. In *Geology Volume IV* (15 p*)*. Oxford: Encyclopaedia of Life Support Systems (EOLSS).

Dawson, J. B. (1964). *Quarter Degree Sheet 54: Monduli.* Tanzanian Geological Survey.

Dawson, J. B. (2008). The Gregory Rift Valley and Neogene-recent volcanoes of northern Tanzania. *Geological Society London Memoir, 33,* 102 p.

Grey, I. M., & Thomas, C. M. (1964). *Quarter Degree Sheets 6 and 14: East Mara.* Mineral Resource Division of Tanzania.

Kasanzu, C., & Manya, S. (2010). Stratigraphic and sedimentological evolution of the neo-Proterozoic Ikorongo Group of north-eastern Tanzania. *South African Journal of Geology, 113,* 361–368.

MacFarlane, A. (1965). *Quarter Degree Sheet 25: Mara.* Mineral Resource Division of Tanzania.

Manya, S., Kobayashi, K., Maboko, M. A. H., & Nakamura, E. (2006). Ion microprobe zircon U-Pb dating of the late Archaean metavolcanics and associated granites of the Musoma-Mara Greenstone Belt, north-eastern Tanzania: Implications for the geological evolution of the Tanzania Craton. *Journal of African Earth Science, 45,* 355–366.

Pickering, R. (1964). *Quarter Degree Sheet 52: Endulen.* Geological Survey of Tanzania.

Abstract

Mount Elgon (4,321 m) is a giant, free-standing mountain located on the Western Rift Platform of the Gregory Rift Valley. Large parts of the mountain are protected in two national parks, one located in Uganda and one in Kenya. Elgon is an extinct dome-shaped shield volcanic cone with an extensive summit plateau. The latter is disrupted by a near-circular depression which has been interpreted as the eroded remnants of a caldera. The Early Miocene age (22 Ma), which makes Elgon one of the oldest volcanoes in the region has resulted in the cone being deeply eroded. The cone is built directly on Basement terranes, an unusual feature as most of the volcanoes associated with the Gregory Rift are constructed on older plateau-style volcanics. The average composition of the Elgon Volcano is nephelinite, typical of the group of magmas associated with the pre-rift stage of the East African Rift System (EARS). Details of the eruptive history are sparse although the cone is probably dominated by tephra and agglomerate with only very minor lava flows. This is typical of a volcano with an extremely high explosive index. The presence of hot springs is indicative of minor geothermal activity. Possibly the most interesting feature of Elgon is the occurrence of small caves, including the famous Kitum Cave, which are regularly accessed by large game including elephant, seeking deposits of salt. The caves are erosional features, rather than lava tubes. The central slopes of Mount Elgon are thickly forested and the mountain supports a remarkable ecosystem that includes numerous endemic species of plants and flowers. The volcanic rocks are deeply weathered and severe landslides are a major hazard, notably on the Ugandan side where human encroachment onto the lower slopes and the removal of forest has caused considerable problems.

Keywords

Botanic zones • Kitum Cave • Landslides
Mount Elgon • Nephelenite • Pre-rift volcanism

Photographs not otherwise referenced are by the author.

© Springer International Publishing AG, part of Springer Nature 2018
R. N. Scoon, *Geology of National Parks of Central/Southern Kenya and Northern Tanzania*, https://doi.org/10.1007/978-3-319-73785-0_8

Plate 8.1 The saddle leading to Wagagai Peak on the Ugandan side of Mount Elgon separates the caldera (left), from the outer slopes (right)

8.1 Introduction

The free-standing Mount Elgon Volcano is located on the regional plateau north of Lake Victoria and cuts across the international boundary between Uganda and Kenya (Fig. 8.1). The dome-shaped profile with a gentle slope (only 4°) is typical of a deeply eroded shield volcano. With a diameter of 80 km, surface area of over 3,000 km² and elevation of almost 3,000 m above the regional plateau, Elgon is one of the largest volcanic centres in East Africa. The highest peak of Wagagai (4,321 m) is situated on the rim of a partially-collapsed caldera on the Ugandan side of the mountain (Plate 8.1), although the summit plateau is as equally extensive and almost as high on the Kenyan side (Fig. 8.2).

Elgon is an important regional water resource for the region. The Suam River drains northward into Lake Turkana whilst the Nzoia and Lwakhakha Rivers flow southward into Lake Victoria. The central slopes of the mountain are thickly forested with the upper slopes reporting extensive heath and moorlands (Plate 8.2a, b). Large areas are protected in national parks on both the Ugandan and Kenyan side. Both parks offer extensive hiking trails. The parks are generally approached from either Mbale in Uganda, or Kitale in Kenya. The 3–4 day hike to the Wagagai Peak via the Budaderi gate, Uganda, provides an opportunity to examine the various botanical zones as well as some important cliff sections and outcrops on the summit plateau. The trek along the Suam Gorge reveals spectacular scenery. The caves located on the Kenyan side are an additional highlight of a visit.

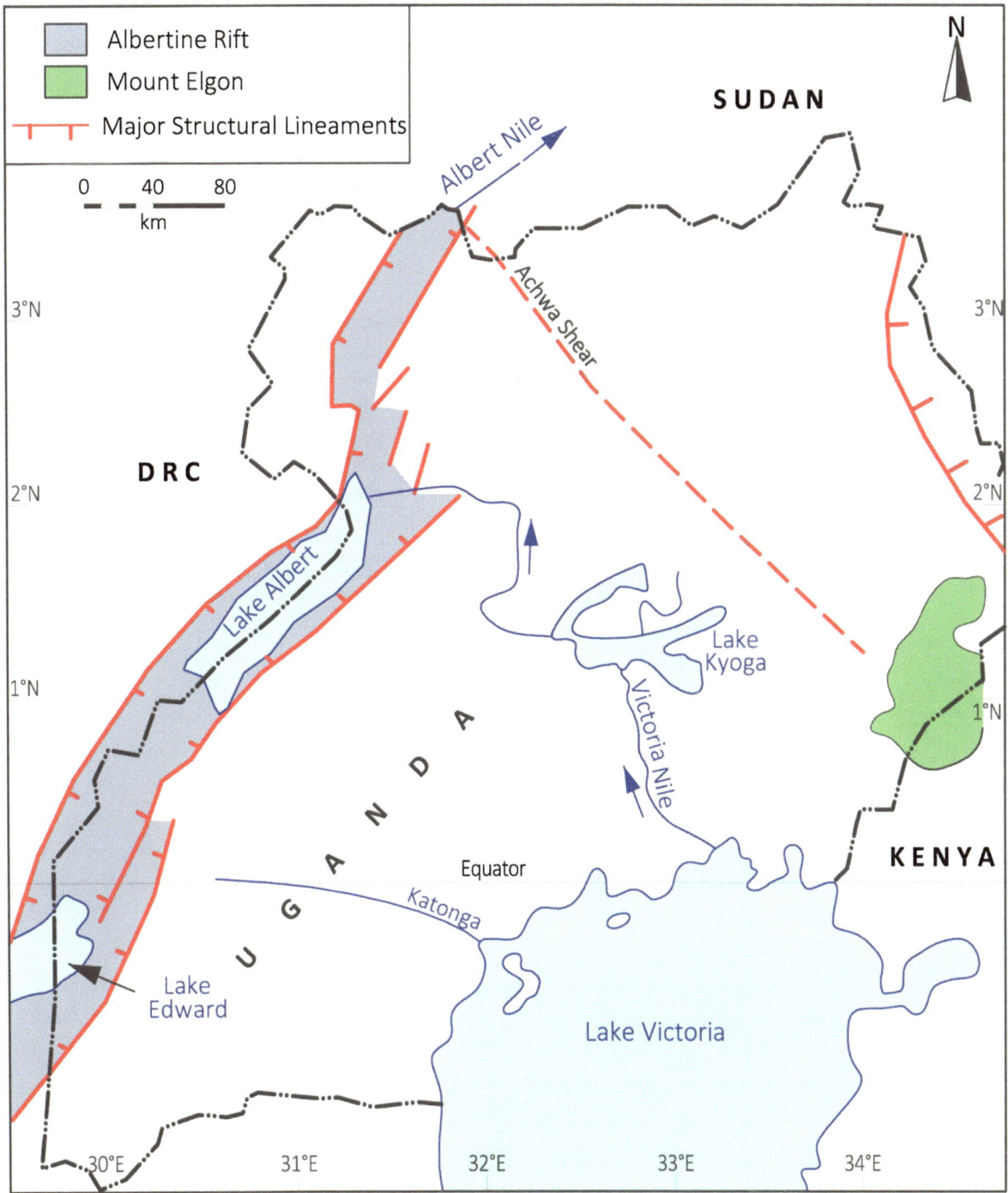

Fig. 8.1 Location map of the Elgon Volcano relative to Uganda and the Albertine Rift

8.2 Pre-rift Volcanism

Mount Elgon is the only example of volcanism related to the pre-rift stage of the EARS that has a sufficiently large cone to be preserved in a park or conservation area. The Early Miocene age (22 Ma) makes Elgon one of the oldest volcanoes in the region. The location on the Western Rift Platform of the Gregory Rift is significant because as rifting developed to the half- and full graben stages, volcanism shifted both southward and onto the Eastern Rift Platform.

8.3   Basement

The Elgon Volcano is built directly on the basement complexes (Davies 1952), a rather unusual feature as most cones in the Gregory Rift are developed on much younger, rift-related volcanic terranes (Fig. 8.3). This feature is, however, consistent with the relative age of Elgon. The western sector of the volcano overlies granite-gneiss of the Central African Craton, part of a more-or-less contiguous block between western Kenya and the Albertine Rift in Uganda/DRC. The eastern sector of the mountain is located to the east of the craton, on the Neoproterozoic Mozambique Belt. Magma may have been fed through a central conduit that exploited a structural weakness where the Achwa Shear, a major (and ancient) structural lineament in Uganda (Fig. 8.1) intersected the eastern edge of the Central African Craton (Fig. 5.2). Localised parts of Mount Elgon overlie Miocene age sedimentary rocks and volcanics. The lahars at the base of the mountain on the southern and southwestern sides are thought to be related to an early phase of activity associated with Elgon. The region also includes a number of pre-Elgon (Miocene) intrusive bodies, including carbonatites in the Bududa District, e.g., Budeda Hill (King et al. 1972). An important cache of Miocene fossils, including mammals, aquatic reptiles, birds, and crabs, has been discovered in sedimentary rocks near Butwa on the northeastern slopes of Mount Elgon (Bishop et al. 1969).

8.4   Volcanology

The Mount Elgon Volcano was first mapped and described in detail by Davies (1952). The volcano is dominated by nephelinitic tephra and agglomerates (Plate 8.3a, b), typical of the Group (i) magmas defined by Baker (1987) and described in Sect. 5.9. The clasts in the agglomerates are

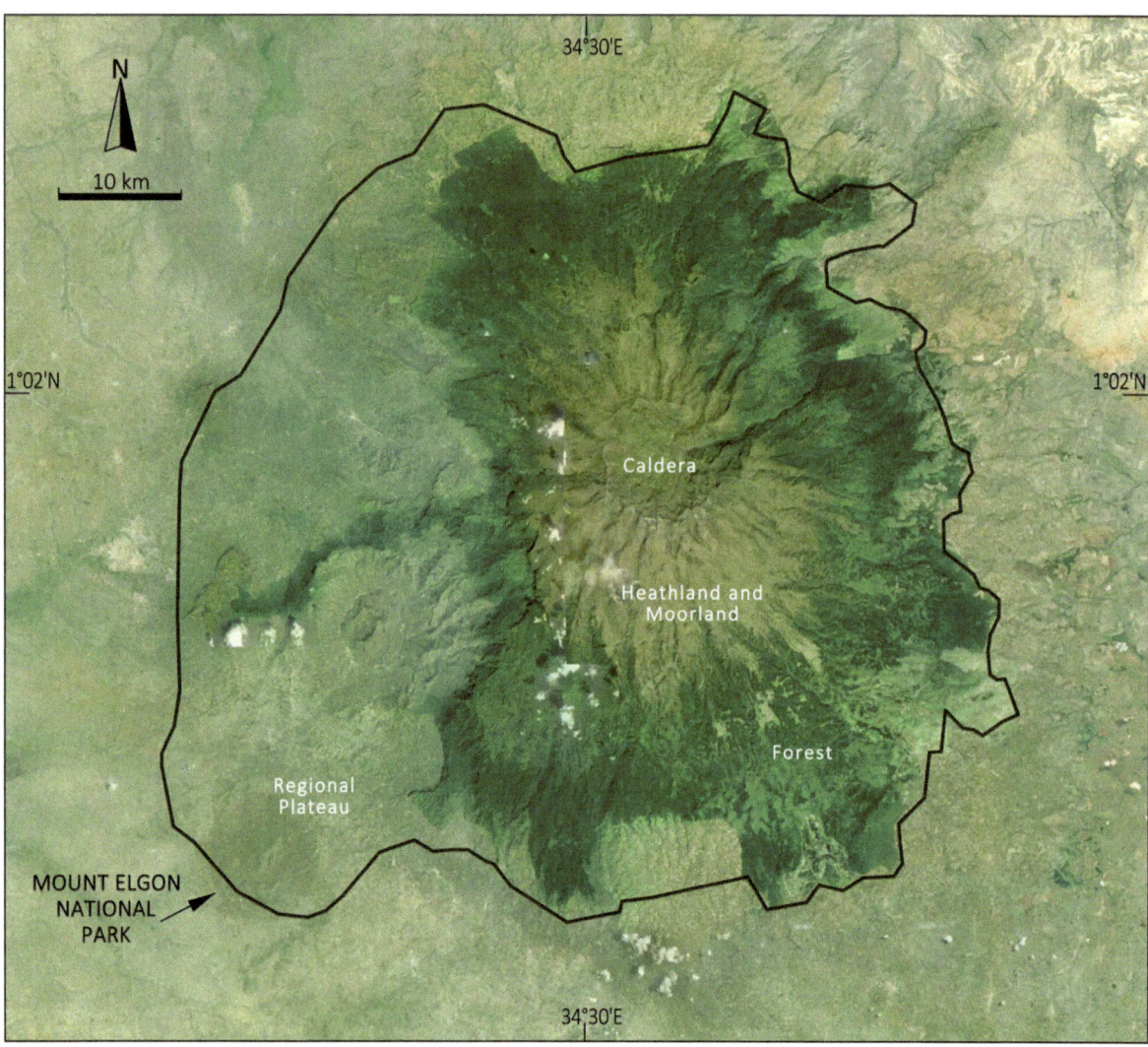

Fig. 8.2 Mount Elgon reveals extensive montane forests on the central slopes (dark green) and in the central depression (probably a caldera) of the summit plateau. *Source* Google-Earth Image 2017, Landsat/Copernicus

typically <10 cm, although locally blocks of nephelinite and phonolite up to 1 m in size occur. Well-bedded tuffs are a subordinate feature, but lavas (nephelinite and minor phonolite and trachyte) contribute less than one per cent by volume of the volcanic pile. The predominance of tephra in comparison to lavas is indicative of a high explosive index. The age of 22 Ma has been ascertained by radiometric dating of nephelinite lava (Bishop et al. 1969; Simonetti and Bell 1995), but whether this is representative of the tephra is uncertain. In addition to the lavas, small intrusive bodies (plugs, dykes, and sills) of alkaline composition are widely dispersed on the volcano. Some of these intrusive bodies contain deposits of apatite and zircon, and possibly rare-earth elements, although they have not been found in economic quantities (Woolley 2001).

8.5 Summit Plateau

Mount Elgon is capped by an extensive summit plateau with a diameter of 8 by 6 km. An interior depression within the plateau has been interpreted as a caldera (Ödman 1931), although this feature was not recognised by Davies (1952). Faults have not been identified on the perimeter of the depression, although a satellite image does suggest evidence of linear structures (Fig. 8.2). Five peaks of over 4,000 m occur on the summit plateau. The highest peak of Wagagai, as well as Little Wagagai (4,298 m), occurs in Uganda, with Lower Elgon (4,301 m), Sudek (4,176 m) and Koitoboss (4,187 m) on the Kenyan side of the mountain. Large moraines, indicative of Late Pleistocene glaciations have been reported from the summit plateau.

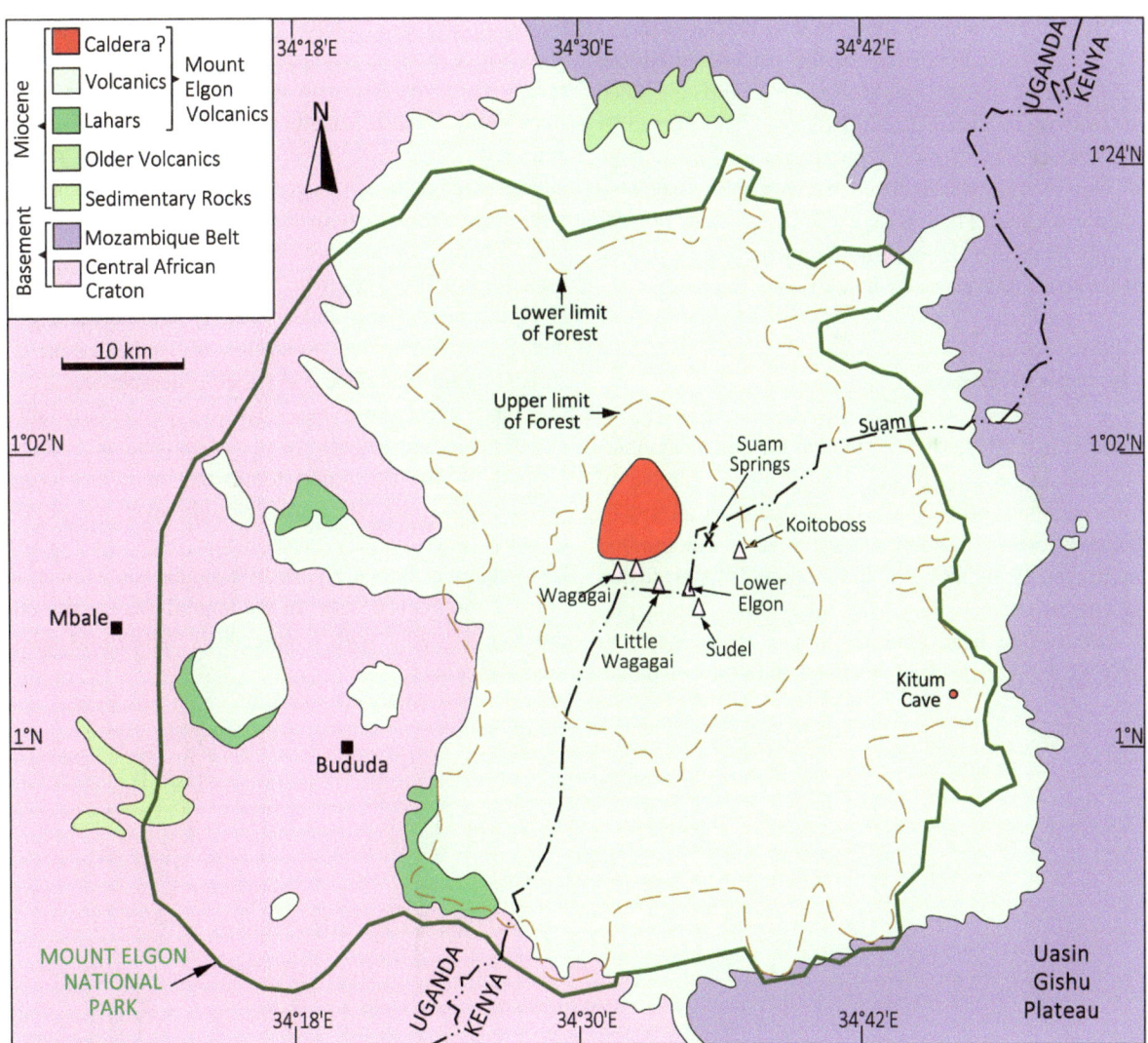

Fig. 8.3 Geological map of Mount Elgon simplified from various sources including Woolley (2001). The Miocene age volcano, an example of the pre-Rift volcanism associated with the EARS, is located on the eastern edge of the Central African Craton

8.6 Hot Springs and Caves

There are several hot springs on Mount Elgon, most of which occur on the eastern slopes, notably in the Suam Gorge on the Kenyan side. Mount Elgon is particularly well known for the occurrence of voluminous caves up to 250 m in length. The caves occur in pyroclastic rocks and contain sodium-rich salt deposits sought by large wildlife. Elephants (*Loxodonta africana*) regularly enter caves to gouge the walls. The "elephant caves" as they are known were first described by Thomson (1880) and they may have provided inspiration for part of Rider Haggard's novel *King Solomon's Mines*. The Kitum Cave is internationally known after a publication by Richard Preston in 1994 disclosed that two visitors died after contracting the rare Marburg Virus. There is considerable debate as to whether the caves are natural (dissolution percolation) or zoogeomorphic (human or elephant). Some reports suggest they are lava tubes but this is unlikely in light of composition of the volcano (predominance of tephra). Lundberg and McFarlane (2006) concluded that the caves are relatively recent (Holocene) and formed by processes induced by erosion. The erosion of incompetent beds of agglomerate was initiated by waterfalls associated with surface streams. There is no evidence of water flowing in the caves, or of dissolution. Initial undercuts are enlarged by roof collapse, as well as by humans and elephants.

8.7 Botanical Zones

Mount Elgon is particularly well known for botanical zones that include a large array of trees and plants, a number of which are endemic to the mountain (Williams et al. 1967). The montane forest is notably extensive, as is the bamboo belt that fringes the upper heath and moorlands. A profusion of wild flowers occur within these zones. The summit plateau is part of the heath and moorlands (Plate 8.4a), as proximity to the Equator and relatively modest height are such that the Alpine deserts and icefields that characterise Mount Kenya and Kilimanjaro are not developed. The area within the summit depression has the appearance of a "Lost World" as it includes stands of giant groundsel (*Senecio barbatipes*) and lobelia (*Lobelia elgonensis*).

8.8 Landslides

The western slopes on the Ugandan side of Mount Elgon are intensively settled and provide a relatively high level of both subsistence farming and cash crops, the latter including good quality coffee (Plate 8.4b). The local people have long depended on the forested slopes of Elgon for part of their livelihood and this has resulted in some conflict with the national park. In recent years there has been a relaxation of rules in some sections and harvesting of resources such as bamboo poles and bamboo shoots (a local delicacy) is now allowed. A series of devastating landslides have occurred in recent years. In 2010, an estimated 350 people lost their lives in the Bududa District after several landslides, and on June 25th 2012, after several days of torrential rain, landslides buried parts of villages on the western slopes of Elgon with considerable loss of life (Plate 8.4c). Landslides have also occurred on some of the pre-Elgon intrusive bodies, including carbonatites where fertile soils encourage intensive farming on steep slopes. A study to understand land use changes including the possibility of landslides on critical slopes, found that since 1933 a considerable loss of woodland and forest cover had occurred, particularly on steep slopes (National Environmental Management Authority of Uganda 2010). The report found that poor land usage (in part attributed to unacceptably high levels of population growth) was largely responsible for the landslides. Encroachment into the park in Uganda has another unfortunate consequence: most of the large fauna has migrated to the Kenyan side, with a potentially deleterious effect on tourist revenues.

Plate 8.2 **a** The giant, dome-shaped mass of Mount Elgon rises above the forested lower slopes of the regional plateau on the Uganda side; **b** Heathlands and moorlands occupy the high slopes of Mount Elgon, near the rim of the caldera, Uganda

(a)

(b)

Plate 8.3 **a** Cliffs of light-coloured nephelinite ashes can be observed on the lower slopes of Mount Elgon, Uganda; **b** Nephelinite agglomerate with medium and small clasts, Mount Elgon, Uganda

Plate 8.4 **a** Heathlands with giant groundsel and lobelia occur in the central caldera of Mount Elgon, Uganda; **b** Lower western slopes of Mount Elgon, Uganda, are intensively farmed in an area of high rainfall and nutrient-rich volcanic soils; **c** Landslide in the Bududa District on the lower slopes of Mount Elgon. *Source* National Environmental Management Authority of Uganda (2010)

Plate 8.4　(continued)

References

Baker, B. H. (1987). Outline of the petrology of the Kenyan Rift alkaline province. In J. G. Fitton & B. G. J. Upton (Eds.), *Alkaline igneous rocks* (Vol. 30, pp. 293–311). Geological Society of London Special Publication.

Bishop, W. W., Miller, J. A., & Fitch, F. J. (1969). New potassium-argon determination relevant to the Miocene fossil mammal localities in East Africa. *American Journal of Science, 267,* 669–699.

Davies, K. A. (1952). *The building of Mount Elgon, East Africa* (Memoir 8, 76 p). Geological Survey of Uganda.

King, B. C., Le Bas, M. J., & Sutherland, D. S. (1972). The history of the alkaline igneous complexes of eastern Uganda and Western Kenya. *Journal of the Geological Society of London, 128,* 173–205.

Lundberg, J., & McFarlane, D. A. (2006). Speleogenesis of the Mount Elgon elephant caves, Kenya. *Geological Society of America Special Paper, 404,* 51–63.

National Environmental Management Authority of Uganda. (2010). *Landslides in Bududa District: Their causes and consequences,* 16 p.

Ödman, O. H. (1931). Volcanic rocks of Mount Elgon in British East Africa. *Geologiska Foreningrens I Stockholm Forhandlingar, 52,* 455–537.

Simonetti, A., & Bell, K. (1995). Nd, Pb, and Sr isotopic data from the Mount Elgon volcano, eastern Uganda-Western Kenya: Implications for the origin and evolution of nephelinite lavas. *Lithos, 36,* 141–153.

Thomson, J. (1880). Notes on the geology of East-Central Africa. *Nature, 28,* 102–104.

Williams, J. G., Arlott, N., & Fennessy, R. (1967). *Collins field guide to National Parks of East Africa* (336 p). Hong Kong: Harper Collins.

Woolley, A. (2001). *Alkaline rocks and carbonatites of the world. Part 3: Africa* (372 p). Geological Society of London.

Aberdare and Mount Kenya National Parks

Abstract

The Aberdare and Mount Kenya National Parks are upland areas located on the Eastern Rift Platform to the north of Nairobi, central Kenya. The dominant physiographic features of the two parks is the Aberdare Range, a 160 km belt of rolling hills and high plateaus, and Mount Kenya, a massive free-standing mountain with a diameter of more than 100 km. The Aberdare Range is comprised of Late Miocene age basaltic volcanics that flooded and smoothed out the topography of this section of the rift platform. Two volcanic groups are identified, the older Satima Volcanics, which dominate the southern part, and the younger Simbara Volcanics, located in the central and northern parts. The dome-shape of Mount Kenya is characteristic of a giant shield volcano. The gentle outer slopes rise some 4,000 m above the regional plateau to an iconic group of central peaks. The volcanism is dominated by the Mount Kenya Volcanic Suite, a thick succession of Pliocene age basalt and phonolite lavas erupted from a central conduit. Blocking of the conduit at approximately 2.6 Ma by an intrusive plug of syenite marked the end of the main phase of activity. The syenite plug is associated with most of the central peaks, including the two highest of Batian (5,199 m) and Nelion (5,188 m). These are formidable obstacles that can only be climbed with a degree of technical difficulty; most trekkers settle for the gentle slopes of Point Lenana (4,985 m). Mount Kenya was severely glaciated during the Late Pleistocene Ice Ages, but the small flank glaciers that surround the central peaks are relics of a Holocene age icecap that was far larger when first observed by Europeans in the latter part of the nineteenth century. Large parts of the Aberdare Range, as well as the lower slopes of Mount Kenya are covered by montane forests that are unusually luxuriant due to the nutrient-rich volcanic soils and proximity to the Equator. These forests are famous for the occurrence of unusual melanistic variants of large mammals, including leopards.

Keywords

Central peaks • Flank glaciers • Melanistic mammals Montane forest • Plateau-style volcanism • Shield volcano

Photographs not otherwise referenced are by the author.

Plate 9.1 The Naro Moru entrance gate to the Mount Kenya National Park provides a view of the gentle outer slopes that characterise the bulk of the volcano

9.1 Introduction

The Aberdare Range and Mount Kenya are part of the Central Highlands, an extensive area on the Eastern Rift Platform of the Gregory Rift, central Kenya. Both areas are covered by large, well-known national parks. The Aberdare National Park was established in 1950 and made famous from visits by the British royalty in the early 1950s, including one to the unique Treetops Lodge. The Mount Kenya National Park was established in 1949 and has recently been proclaimed a World Heritage Site (Plate 9.1). Mount Kenya is one of the iconic images of one of the most scenic countries in the world. Both national parks are served by regional towns that include Nyeri and Naro Moru. The climate of the entire region is temperate and mist is a regular occurrence on higher ground, despite the equatorial setting. Both parks are dominated by volcanic rocks associated with the East African Rift System (EARS). The Aberdare Range is part of an extensive, faulted volcanic terrane derived from numerous fissure eruptions which occurred on the platform to the east of the Gregory Rift. Mount Kenya is a giant shield volcano built around a central conduit.

Mount Kenya is one of Africa's significant mountaineering challenges and the volcanic origin and presence of ice sheets, first observed by Joseph Thomson in 1883 (Thomson 1887), was only accepted by European geographers when J. W. Gregory reached the Lewis Glacier at a height of approximately 5,000 m in 1893 (Gregory 1894a). Gregory also reported on the volcanic origin of this huge, free-standing mountain. The central peaks, including the two highest of Batian (5,199 m) and Nelion (5,188 m) were first climbed by H. J. Mackinder in 1899, who ascertained they are part of a large intrusive plug of syenite.

9.2 Aberdare National Park

The Aberdare National Park has an area of 766 km^2 and covers large sections of the 160 km long, north–south trending Aberdare Range. The western boundary is demarcated by the Satima Fault, which forms a huge escarpment on the eastern boundary of the Gregory Rift (Fig. 9.1). The southern part of the Aberdare Range is separated from the rift valley by the Kinangop Plateau, a step-like platform bounded by faults downthrown to the west. The eastern boundary of the Aberdare Range is less well-defined as it

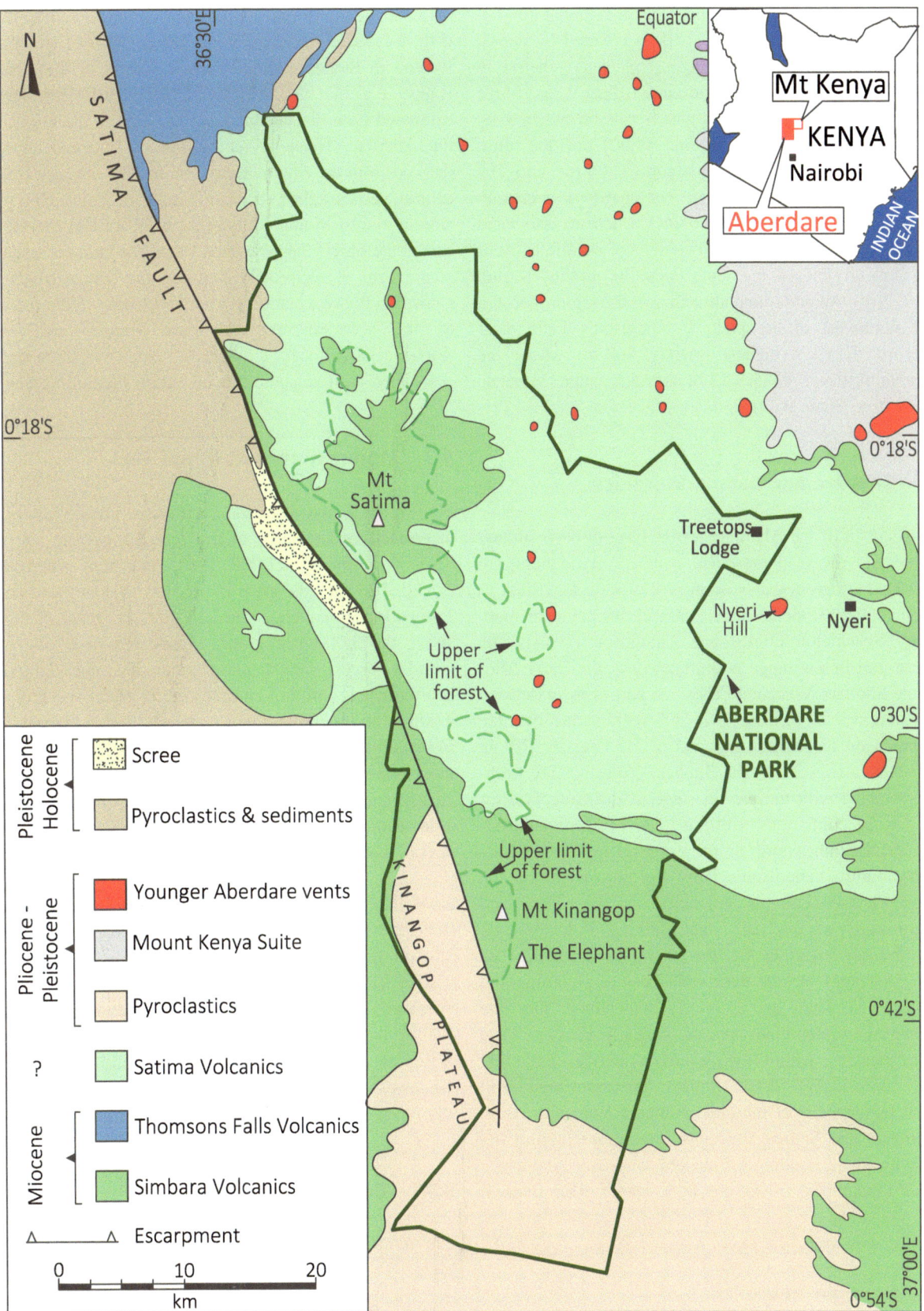

Fig. 9.1 Geological map of the Aberdare National Park simplified from the Kijabe Quarter Degree Sheet 43 (southwest) and Nyeri Quarter Degree Sheet 43 (northeast)

merges into a broad valley that includes the regional town of Nyeri. The bulk of the national park is covered by dense, montane forests with only the highest plateaus penetrating into the zone of heath and moorlands (Plate 9.2a). Some plateaus reveal broad summits which can be reached by full-day treks, e.g. Mount Satima (4,001 m), the third highest mountain in Kenya, Kinangop (3,906 m), and The Elephant (3,600 m). The Aberdare National Park is characterised by deep, thickly forested ravines which are associated with a myriad of watercourses. Several large waterfalls occur on the rims of plateaus, e.g. Gura Falls (302 m) and Karura Falls (272 m). An additional attraction of the region, located to the northwest of the park, is Thomson's Falls, near Nyahururu, Kenya's highest town (2,360 m) where the Ewaso Naruk River plunges 72 m over a resistant, Miocene age lava flow (Plate 9.2b).

9.2.1 Simbara and Satima Volcanics

The volcanism of the Aberdare Range has not been studied in detail and the map presented here is very generalised (Fig. 9.1). Two groups of volcanic rocks were identified as the principal constituents of this area during regional mapping by Shackleton (1945) and Thompson (1964). The southern part of the range is dominated by the Early Miocene age Simbara Volcanics. They are part of the extensive faulted terranes of basaltic lava and agglomerate that are widespread on the eastern margins of the Gregory Rift in central Kenya (Fig. 5.2). The Simbara Volcanics are probably associated with the pre-rift stage of the Gregory Rift. They may be contiguous with the Early Miocene age volcanics located father north on the Eastern Rift Platform, i.e. near Nyahururu (Thomson's Falls Volcanics) and on the Laikipia Plateau (Laikipia Volcanics). In the central and northern parts of the park, the Simbara Volcanics are unconformably overlain by the Satima Volcanics. The Satima Volcanics include thick sequences of phonolite and trachyte lavas. Their age is poorly constrained: they are thought to be either Late Miocene or Early Pliocene. The highest plateaus within the Aberdare Range, including (rather confusingly) Mount Satima, are associated with uplifted fault blocks of the older Simbara Volcanics. Both the Simbara and Satima Volcanics were erupted from fissures with lavas flooding an ancient landscape to form a stepped topography of rolling hills, rather than discrete cones.

9.2.2 Younger Volcanics

Pyroclastic rocks located to the south of the Aberdare Range are associated with Plio-Pleistocene eruptions either in the rift valley, e.g., Mount Longonot or on the Kinangop Plateau (Fig. 9.1). The Aberdare Vents are a group of small cones located in the northern and central parts of the Aberdare Range, e.g. the basaltic lava at Nyeri Hill. They have an estimated Late Pleistocene age. The area to the northeast of the Aberdare Range is comprised of Pliocene age lavas derived from the Mount Kenya Suite. The absence from the central parts of the Aberdare Range of volcanism with a similar age to Mount Kenya (the Simbara and Satima Volcanics are older; the Aberdare Vents are younger), suggests some form of reciprocity between the source of the volcanism in the two centres. Pyroclastic rocks in the rift valley to the west of the Aberdare Range are younger (Pleistocene-Holocene), as are the extensive scree deposits associated with episodic movement on the Satima Fault.

9.3 Mount Kenya National Park

The Mount Kenya National Park was originally established to protect the landscape of this iconic mountain above the 3,364 m contour (11,000 feet), but was subsequently extended to include some of the lower slopes and forests. The park now covers some 715 km², i.e. most of the area above the 3,000 m contour. Mount Kenya is one of the world's largest free-standing mountains: the diameter is approximately 100 km and the central peaks rise more than 4,000 m above the regional plateau. Prior to extensive erosion during the Pleistocene, it has been estimated the mountain included a summit crater with an elevation of over 7,000 m. The main bulk of Mount Kenya resembles a giant dome-shape typical of shield volcanoes. The group of central peaks constitutes only a small fraction of the overall massif (Fig. 9.2). Mount Kenya is the catchment for the Tana and Eyaso Ng'iso Rivers, which drain the eastern and western slopes, respectively. Large numbers of people in the region rely on the mountain for water and there are numerous smaller streams. The mountain has created an area of increased rainfall in comparison to the drier regional plateaus due to the orographic effect of the southeast monsoon. A typical, daily weather pattern is the build-up of clouds on the lower western slopes in the late morning, which generally cover the central peaks by early afternoon.

9.3.1 Central Peaks

The two highest peaks of Batian and Nelion, together with Midget (4,770 m) and Point John (4,883 m) have distinctive, triangular shapes, whilst most of the subsidiary peaks, including the Gendarmes are part of the northern, southern and western summit ridges (Plate 9.3a). Batian and Nelion are separated by a 250 m wide/140 m deep gap, the Gate of the Mists. Mackinder's pioneering route used the south ridge to reach Batian by this gap. The Petit Gendarme and Great Gendarme are obstacles on the west ridge of Batian, first climbed by E. G. Shipton and H. W. Tilman in 1930. Most of the central peaks can be ascended only with a degree of technical difficulty. Trekkers generally aim for the uppermost part of the gentle slopes of Point Lenana (4,985 m). The most popular route is the Naro Moru track which takes 2 or 3 days via the Teleki Valley with overnights at the Meteorological Station and Mackinder Huts (Plate 9.3b).

9.3.2 Mount Kenya Volcanic Suite

The Mount Kenya Volcano was mapped in considerable detail by Baker (1967). The bulk of the mountain is comprised of Pliocene age components of the Mount Kenya Volcanic Suite (Fig. 9.2). The maximum age is 4.5 Ma. Some minor activity may have persisted into the Early Pleistocene (Baker 1987; Woolley 2001). The Mount Kenya volcanics unconformably overlie gneiss and schist of the Neoproterozoic Mozambique Belt in an area to the south of

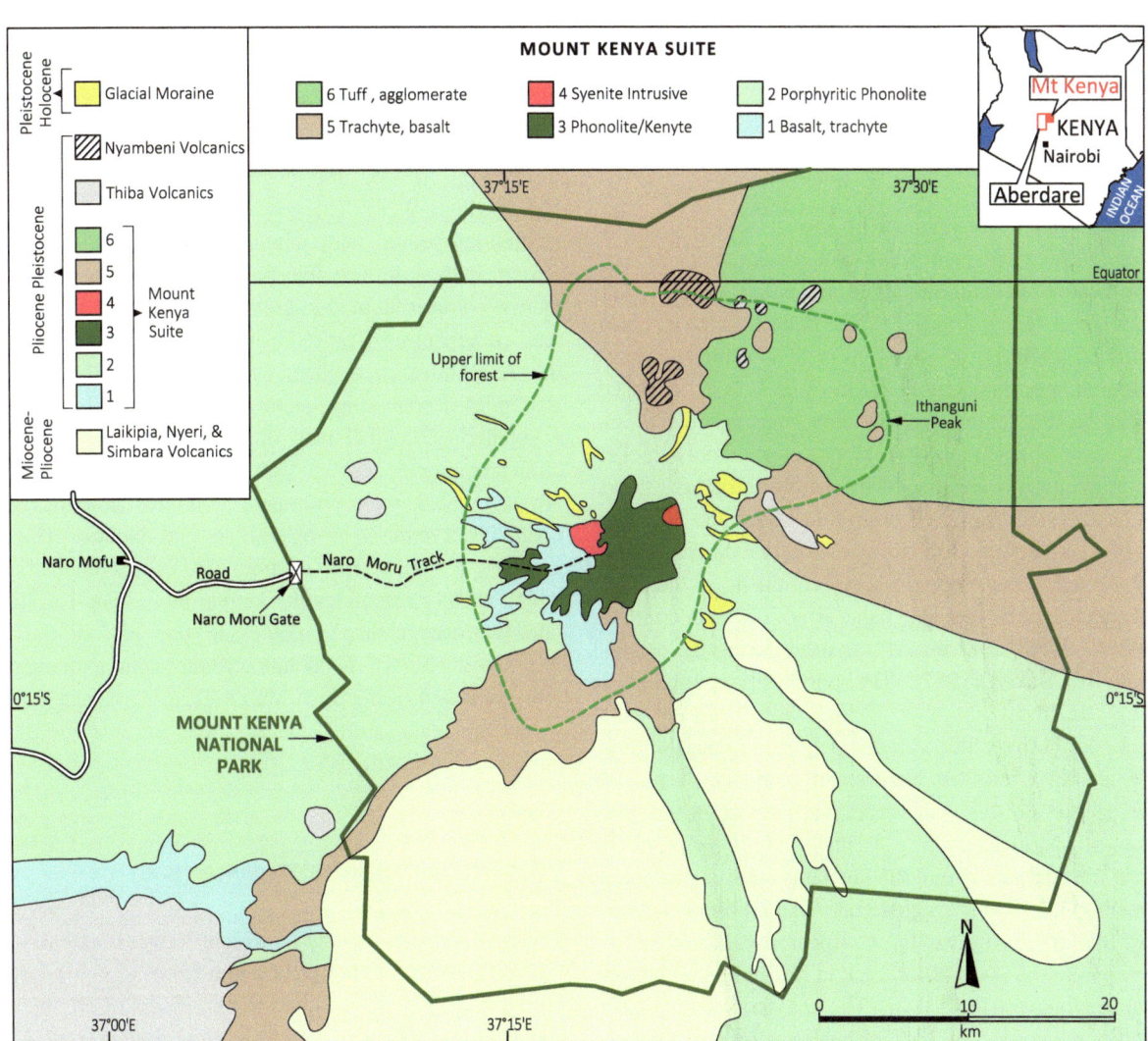

Fig. 9.2 Geological map of the Mount Kenya National Park simplified from the Mount Kenya Quarter Degree Sheet 44 (southwest) and Baker (1967). The bulk of the volcano is comprised of lavas and agglomerates, but the central peaks correlate with one of several intrusive bodies of nephelinite syenite

the map. To the west, the volcanic suite overlies the extensive volcanic terranes of Early and Late Miocene age located on the Eastern Rift Platform. The Nyambeni Hills Volcanics to the northeast of Mount Kenya report a wide range in ages (4.5–0.46 Ma), but in general are probably younger than Mount Kenya (Woolley 2001). Baker (1967) recognised three extrusive events and one intrusive event within the Mount Kenya Volcanic Suite. An early group of parasitic events (not shown on the map), included flows of basalt on the lower southwestern slopes and trachyte on the higher southern slopes. Details of the main volcanic component (group 1–3 on Fig. 9.2), the intrusive event (group 4 on Fig. 9.2), and the later parasitic activity (groups 5–6 on Fig. 9.2) are reported below.

9.3.3 Main Volcanic Component (1–3)

The main volcanic component of Mount Kenya is a succession of extensive lava flows that yield radial, outward dips indicative of eruption from a central conduit. The exposed flows (excluding the earlier parasitic activity) have a combined thickness of over 600 m. It is estimated that a thickness of approximately 2,850 m of lavas is not exposed. Proximal to the central conduit, flows are relatively thin, with those on the gentle, lower slopes considerably thicker. Three broad groups are recognised within these lava flows: basalt and trachyte (1), porphyritic phonolite (2) and phonolite and kenyte (3). Kenyte describes a distinctive glassy phonolite that is particularly prominent on the upper slopes (Gregory 1894b). This name is no longer in common usage but is retained here as it was used during the original mapping by Baker (1967). The porphyritic phonolite is probably the most abundant of the three types of lava, cropping out in low cliffs with distinctive columnar jointing. It weathers into rounded blocks and boulders. A notable feature is the presence of large, partially resorbed phenocrysts of plagioclase, up to 4 cm in length. Some of the feldspar phenocrysts occur as thin laths with a prominent alignment (Plate 9.4a). Agglomerate is a common component of this group of lavas on the upper slopes, and can be observed from the Naro Moru track in low cliffs that include several small caves (Plate 9.4b). The kenyte is distinguished from the phonolite by the presence of phenocrysts of nepheline, as well as by vesicles filled by small, pale-coloured crystals of zeolite. Weathered faces of the kenyte display orange or purple colours. The kenyte is typically intercalated with thick pyroclastic deposits.

The sequences of kenyte and pyroclastics were thought to have been erupted during a phase of volcanism which oscillated between quiescent lava flows and highly explosive tephra events. The latter were associated with intrusive plugs

that temporarily blocked the central conduit and were subsequently ejected as volcanic bombs.

9.3.4 Nepheline Syenite Intrusive (4)

The main phase of volcanism at Mount Kenya was terminated by intrusion of a large plug, which blocked the central conduit, as well as subordinate plugs in the smaller, subsidiary vents (Baker 1967). The intrusive plugs have been dated at 3.1–2.64 Ma. The central plug was interpreted as the last in a sequence of events (earlier plugs were blown out to form pyroclastic deposits). The central plug is a concentrically zoned, cylindrical body comprising a central core and a chilled rim. The overall diameter is approximately 2.4 km and the depth at least 900 m as determined from exposures on the precipitous cliffs of Batian and Nelion (Plate 9.2a). The core of the plug is composed of nepheline syenite, with tabular crystals of feldspar up to 1.5 cm in length. The feldspar crystals are emplaced in a matrix comprised of fine-grained mosaics of feldspar and reddish nepheline. The reddish coloration is pronounced in the Amphitheatre, a sheer face on the northern side of Batian, as described by Jennings (1971) who also published a detailed geological map of the central plug. The steep cliffs and angular sections that characterise the central peaks are associated with prominent columnar jointing of the nepheline syenite e.g. on Batian and Nelion. The plug includes a chilled rim of phonolite, the composition of which closely resembles the kenyte lavas. The rim divides into two wedge-shaped bodies in one part of the plug. A smaller plug of nepheline syenite with a diameter of 450 m located near Polish Man's Tarn is associated with a secondary vent from which some of the porphyritic phonolites were erupted.

9.3.5 Later Parasitic Activity (5–6)

The later parasitic activity on Mount Kenya is extensive, but for sake of clarity only two of a number of groups are shown on the map (Fig. 9.2). The oldest are flows of olivine basalt and trachyte, which have a combined thickness of 470 m. They were erupted from small fissures on the upper, northern and eastern flanks. They form Ithanguni Peak (3,887 m), a prominent parasitic vent. The final phase of activity within the Mount Kenya Volcanic Suite produced cones of lava, agglomerate, and pumice, together with tuffs which are dispersed over a wide area on the northeastern flanks. The parasitic cones show some affinity with the Nyambeni Hills Volcanism with which they may be coeval.

9.3.6 Slope Glaciers

The lower slopes of Mount Kenya show no evidence of glacial activity, but the upper slopes and high peaks were severely glaciated during the Late Pleistocene Ice Ages. The Holocene ice cap originally covered some 400 km². When the early explorers visited the entire summit was ice-capped. The extensiveness of this icecap was reported by Gregory (1894c) and was treated with considerable importance in Europe as it signified that the Ice Ages were global, rather than localised phenomena. The retreat of the ice cap after the Last Glacial Maximum (20,000 BP) was ascertained by Baker (1971) from detailed mapping of moraines. These, mostly linear features are located in the valleys that define a radial pattern at heights of 3,000–4,500 m (Fig. 9.2). A cluster of moraines can be viewed from the Naro Moru track near Teleki Tarn and Shipton Peak.

Twelve individual glaciers were identified by early explorers, of which the Lewis Glacier, on the eastern slopes, was the largest. By the 1980s an area of only 0.7 km² was ice-covered. The Lewis Glacier has been retreating at approximately 7.4 m/annum since 1900. The remnants of the Tyndall, Darwin, and Diamond Glaciers can be observed in the central peaks (Plate 9.3a). Formation and recession of ice sheets on Mount Kenya has probably followed a similar pattern to the other East African Mountains, such as Kilimanjaro.

9.4 Unusual Flora and Fauna

Botanical zones on mountains adjacent to the Gregory Rift, including the Aberdare Range and Mount Kenya, are particularly diverse. The Aberdare Range and Mount Kenya receive higher than average rainfall due to the orographic effect and the equatorial setting. Unusually dense, montane forests have developed on the fertile, nutrient-rich soils that characterise these volcanic terranes. The forests on Mount Kenya occur up to a height of approximately 2,500 m where they are succeeded by zones of bamboo forest (up to 3,000 m), heath (up to 3,500 m), and moorlands (up to 4,500 m). The heath and moorland were described by Gregory (1894a) as having many similarities with the Cape Floral Kingdom. Stands of endemic giant groundsels, e.g. *Senecio brassica* and *Senecio keniodendron* and tree lobelias, e.g. *Lobelia keniensis* and *Lobelia telekii* occur up to heights of 3,800 m (Plate 9.4c).

The Aberdare and Mount Kenya National Parks are well known for the restricted occurrence of rare and endangered species of large mammals, particularly in the montane and bamboo forests, e.g. Bongo (*Boocercus eurycerus*) and Giant Forest Hog (*Hylochoerus meinertzhageni*). Large cats such as Leopard (*Panthera pardus pardus*), Serval (*Felix serval*), and Large Spotted Genet (*Genetta tigrina*) have evolved a melanistic camouflage (e.g. Gregory 1894d; Williams et al. 1967). Males of many of the larger mammal species in the parks, including Elephant, Buffalo (*Syncerus caffer*), and Bushbuck (*Tragelaphus scriptus*) are far darker than normal and may be entirely black. These features can be linked to the rapid speciation (Sect. 1.5). The camouflage probably evolved specifically for the unusually dense nature of the equatorial montane forests of the Aberdare Range, Mount Kenya, and to a lesser extent Mount Elgon. The absence of some of these species and variants from similar habitats, e.g. Kilimanjaro, Mount Meru, and the Ngorongoro Highlands may be due to the less dense nature of the montane forests or the localised nature of Darwinian evolution which appears to have characterised the discrete volcanic terranes of East Africa.

Plate 9.2 **a** Rolling, thickly forested hills of the Aberdare Range give way to heath and moorland on the high summits; **b** Thomson's Falls near Nyahururu plunges some 72 m over Miocene age volcanic lavas

Plate 9.3 a The rocky central peaks of Mount Kenya including the highest points of Batian (5,199 m) and Nelion (5,188 m) are associated with an intrusive body, or plug, of nepheline syenite that blocked the main conduit; **b** The forested lower slopes of Mount Kenya are replaced by heath and moorland at an altitude of approximately 3,800 m. Rocky outcrops reveal exposures of lavas with near-horizontal bedding

Plate 9.4 a The porphyritic phonolite lavas of Mount Kenya may contain closely-packed phenocrysts of plagioclase; **b** Agglomerates with clasts several tens of cm in size indicative of violent eruptions are a common feature on the upper slopes of Mount Kenya; **c** The high moorlands of Mount Kenya include the endemic *Lobelia keniensis*

References

Baker, B. H. (1967). Geology of the Mount Kenya area. *Explanation to degree sheet 44:NW quadrant* (Report 78, 79 p). Geological Survey of Kenya.

Baker, B. H. (1971). The glaciology of Mount Kenya. In J. Mitchell (Ed.), *Guide book to Mount Kenya and Kilimanjaro* (pp. 41–44). Mountain Club of Kenya.

Baker, B. H. (1987). Outline of the petrology of the Kenyan Rift alkaline province. In J. G. Fitton & B. G. J. Upton (Eds.), *Alkaline igneous rocks* (Vol. 30, pp. 293–311). Geological Society of London Special Publication.

Gregory, J. W. (1894a). Mountaineering in Central Africa, with an attempt on Mount Kenya. *Alpine Journal, 17,* 89–104.

Gregory, J. W. (1894b). Contributions to the physical geography of British East Africa. *The Geographical Journal, 4,* 289–315, 408–424, 505–524.

Gregory, J. W. (1894c). Contributions to the geology of British East Africa Part I. The glacial geology of Mount Kenya. *Quarterly Journal of the Geological Society of London, 50,* 515–530.

Gregory, J. W. (1894d). Some factors that have influenced zoological distribution in Africa. *Proceedings of the Zoological Society of London for 1894, 165*–166.

Jennings, D. J. (1971). Geology of Mount Kenya. In: Mitchell, J. (ed.), *Guide Book to Mount Kenya and Kilimajaro.* Mountain Club Kenya, 30–41.

Thompson, A. O. (1964). Geology of the Kijabe area. *Explanation to degree sheet 43: SE quadrant* (Report 67, 57 p). Geological Survey of Kenya.

Thomson, J. (1887). *Through Masailands: A journey of exploration among snow-clad volcanic mountains and strange tribes of eastern Africa* (364 p). London: Sampson Low, Marston, Searle & Rivington.

Shackleton, R. M. (1945). Geology of the Nyeri area. *Explnation to degree sheet 43: NE quadrant* (Report 12, 26 p). Geological Survey of Kenya.

Williams, J. G., Arlott, N., & Fennessy, R. (1967). *Collins field guide to National Parks of East Africa* (336 p). Hong Kong: Harper Collins.

Woolley, A. (2001). *Alkaline rocks and carbonatites of the world. Part 3: Africa* (372 p). Geological Society of London.

Ngorongoro Conservation Area

Abstract

The Ngorongoro Conservation Area (NCA) covers a large, partly mountainous region in northern Tanzania. The diverse ecosystems include the Ngorongoro Highlands with several peaks of over 3,000 m. The most spectacular feature is the Ngorongoro Caldera, a self-contained sanctuary designated as one of the natural wonders of the world. This is one of the premier wildlife destinations in the world, with large concentrations of large mammals occurring on grassy savannahs and in small forests sustained by freshwater lakes and swamps. The caldera is part of the Ngorongoro Volcanic complex, a group of eight, Pliocene age (3.7–1.79 Ma) shield volcanoes associated with the East African Rift System (EARS). This extinct complex of volcanoes is restricted to a discrete structural block, the Eyasi Half-graben, a step-like feature demarcated from the Gregory Rift Valley on the eastern side by a near-vertical regional escarpment. The western boundary of the NCA is associated with smaller, step-like faults on the western edge of the Serengeti Plains. The Ngorongoro volcanism is dominated by alkaline basaltic and trachytic lavas and includes giant calderas resulting from the highly explosive activity which characterises the rift-related activity. The largest feature is the Ngorongoro Volcano (2.5–1.9 Ma) with its almost circular caldera measuring some 22 by 18 km. The caldera is constrained by steep internal walls up to 350 m high. Less well known, but as equally spectacular is the thickly forested Empakaai Caldera located in the north-eastern part of the NCA. Other geological features include exposures of the Neoproterozoic Mozambique Belt in the Gol Mountains and the Shifting Sands near the palaeoanthropological site of Oldupai Gorge. The sand dunes are unusual, isolated features that consist of black volcanic ash derived from Holocene age eruptions of the Oldoinyo Lengai Volcano. The temperate climate of the Ngorongoro Highlands, particularly in comparison to the hot and arid rift valley, together with the nutrient-rich volcanic soils supports extensive farming on the perimeter of the NCA.

Keywords

Caldera • Escarpment • Ngorongoro • Pliocene Sanctuary • Shifting sands • Volcanism

Photographs not otherwise referenced are by the author.

© Springer International Publishing AG, part of Springer Nature 2018
R. N. Scoon, *Geology of National Parks of Central/Southern Kenya and Northern Tanzania*, https://doi.org/10.1007/978-3-319-73785-0_10

Plate 10.1 The lush paradise of the Ngoitokitok Springs in the Ngorongoro Caldera is located beneath the northern caldera wall which in turn is overlooked by the flanks of the Olmoti Volcano

10.1 Introduction

The NCA covers some 8,200 km² of diverse ecosystems that include grassy savannahs, montane forests, heath and moorlands, and small lakes and swamps. The most spectacular component is the Ngorongoro Caldera, one of the natural wonders of the world. This self-contained ecosystem hosts huge herds of wildlife on grassy savannahs and patches of forest that rely on the internal lakes and springs (Plate 10.1). The caldera exhibits remarkable differences between the dry and wet seasons (Plate 10.2). The principal physiographic features of the NCA are the Ngorongoro Highlands, the Salei Plains, part of the Eastern Serengeti Plains, and the remote wilderness area of the Gol Mountains (Fig. 10.1). The Ngorongoro Highlands are asociated with an extinct volcanic complex located wholly within the Eyasi half-graben. The Ngorongoro Volcanic complex includes eight discrete volcanoes of which three reveal large calderas (Box 5.1). Most of the peaks and caldera shoulders rise to heights of 2,500–3,500 m.

The temperate climate of the Ngorongoro Highlands contrasts markedly with the hot, dry and relatively low-lying rift valley. The valley has an elevation of 954 m in the vicinity of Lake Manyara. Whereas the eastern boundary of the NCA is demarcated by a prominent fault (the Western Escarpment of the rift valley), the western boundary is defined by minor faults with modest surface expressions. The palaeoanthropological sites of Oldupai Gorge and Laetoli are located in the NCA, at the foot of the Ngorongoro Highlands on the semi-arid Eastern Serengeti Plains (Chap. 11).

Fig. 10.1 Satellite image of the Ngorongoro Conservation Area (NCA) and the Natron-Manyara branch of the Gregory Rift (approximate width of view 150 km). Most of the NCA is situated in the Eyasi Half-graben, a structural block defined by major and subordinate faults that are invariably downthrown to the east. A major fault demarcates the western boundary of the rift valley. The Oldupai River system (on the Serengeti Plains) is unusual in that it drains eastward, into the Olbalbal Swamps (OS). Note the contrast between the lush, well-vegetated Ngorongoro Highlands in the centre of the image (dark green) and the much drier areas (beige) to the east (rift valley) and west (Serengeti Plains). The southeastern boundary of the Ngorongoro Highlands demarcates the boundary of the NCA with farmland of the Mbulu Plateau (dark brown). Three of the eight centres in the Ngorongoro Volcanic complex include calderas: Empakaai (EC), Ngorongoro (NC), and Olmoti (OC). Four of the volcanoes form prominent cones: Lemagrut (LE), Loolmalasin (LO), Oldeani (OL), and Sadiman (SA). Gelai (GE), Kerimasi (KE), and Oldoinyo Lengai (OD) are younger cones located in the rift valley. Ketumbeine (KB) is a Pliocene age cone located in the divergence. *Source* NASA Landsat 7 ETM + image mosaic for the year 2000 sourced from the University of Maryland Global Land Cover Facility, courtesy of Philip Eales, Planetary Visions

10.2   Approaches

The NCA is generally approached from the eastern side by the road that passes through the village of Mto-Wa-Mbu in the rift valley, north of Lake Manyara (Fig. 10.2). This road climbs steeply up the Western Escarpment i.e. the escarpment on the western side of Lake Manyara prior to levelling off on a plateau to the east of the Ngorongoro Highlands. The ascent provides spectacular views of Lake Manyara (Plate 5.1). The plateau, which includes the regional town of Karatu, is relatively densely populated. The eastern entrance gate to the NCA is located near the base of a second ascent where the road climbs up the thickly forested, southern slopes of the Ngorongoro Volcano. A view point on the rim of the Ngorongoro Caldera reveals the immensity of this famous sanctuary. The road that circles the caldera on the eastern side provides access to the Bulbul Depression and the Empakaai Caldera. The western approach to the NCA is via a road that winds up a long pass from the Eastern Serengeti Plains. The paradox of crossing a graben fault

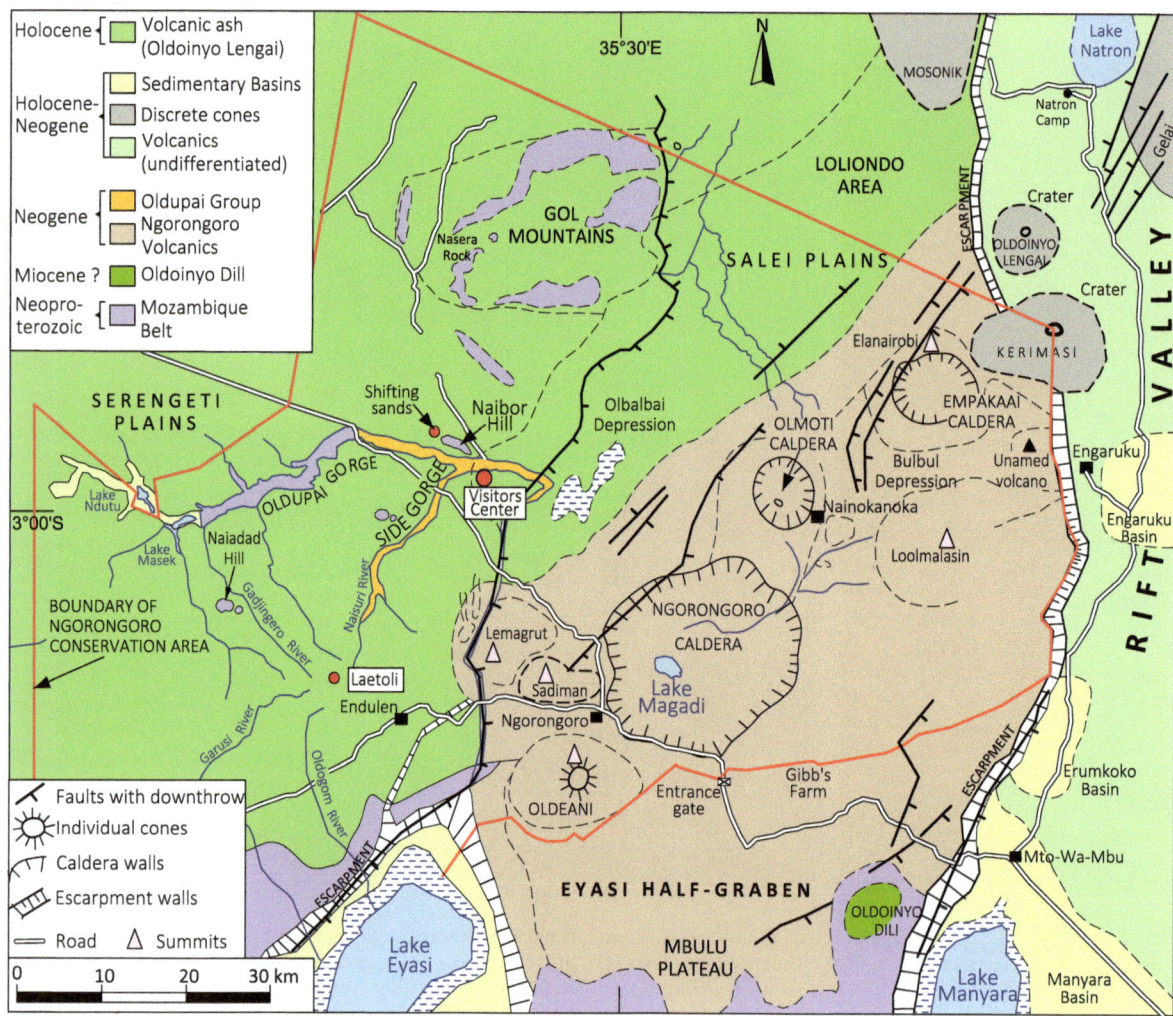

Fig. 10.2   Map of the Ngorongoro Conservation Area simplified from the geological mapping of Pickering (1958, 1964, 1965), Orridge (1965), and Dawson (2008)

downthrown to the east (the western boundary to the Eyasi Half-graben) and yet gaining altitude is reconciled with the great thickness of lavas and ashes associated with the Ngorongoro Volcanic complex.

10.3 Ecosystems

The eastern and southern slopes of the cones and calderas in the Ngorongoro Highlands are generally covered by extensive montane forests, as are the internal walls of some calderas (Fig. 10.1). The two highest calderas, Ngorongoro and Olmoti, are floored by grassy savannahs. The forest on some of the highest caldera rims is displaced by heath and moorlands and some of the high peaks include localised areas of bamboo forest. The drainage includes streams that flow eastward into the rift valley, as well as internally within calderas. The ash and lavas, together with increased rainfall associated with the higher topography, have resulted in thick red, highly fertile soils on the eastern and southern slopes (Plate 10.3a). The area on the eastern perimeter of the NCA is intensively farmed in comparison to the drier plateaus and relatively barren northern and western slopes. Crops such as high-quality coffee are grown, both on shambas and commercially such as at Gibbs Farm, which also caters for tourists. Large parts of the NCA are traditional lands of the Maasai who have ancient rites to inhabit and graze cattle in the conservation area.

10.4 Geological Framework

A simplified geological map illustrates the structural features of the NCA, the bulk of which is situated within the Eyasi Half-graben (Fig. 10.2). The step-like nature of this feature is evident from the observation that the major faults, including those associated with the Lake Eyasi Escarpment and Western Escarpment are downthrown to the east. The Salei Plains constitutes a relatively low-lying block to the north of the Ngorongoro Highlands in an area of subsidiary faulting. The Olbalbal Depression is associated with a small graben bordered by a secondary fault downthrown to the west. This fault is significant as it has blocked the eastward-flowing Oldupai River (Chap. 11). The Gol Mountains are inselbergs located on the Eastern Serengeti Plains to the west of the Eyasi half-graben.

Four principal geological terranes are identified in the NCA. The oldest is the Mozambique Belt (Neoproterozoic)

but the most extensive is the Pliocene age Ngorongoro Volcanic complex. Sedimentary basins associated with the EARS occur in the vicinity of Lakes Eyasi and Manyara. The Oldupai Group crops out in gorges on the Eastern Serengeti Plains and the Laetoli Group (not shown for reasons of scale on the geological map) is associated with an area of badlands erosion on the regional plateau (Fig. 10.1). The youngest feature of the NCA is the Holocene age deposits of ash derived from the Oldoinyo Lengai Volcano.

10.5 The Mozambique Belt

The most prominent of the inselbergs on the Eastern Serengeti plains are the Gol Mountains (or Oldoinyo Ogol). They attain a maximum height of 2,100 m. Smaller features include the Nasera, Naiadad, and Naibor Hills. All of these features are comprised of the Mozambique Belt with an estimated age of 800–500 Ma. The dominant lithology is quartzite and mica schist. The bedding in some of the exposures of quartzite and schist is orientated nearly vertical, typical of mobile belts that have been severely deformed, e.g., Ol Karien Gorge a spectacular locality where the near-vertical sidewalls enclose a narrow slot. The Nasera Hill is a classic monolith located some 27 km north of Oldupai Gorge which rises to a height of 350 m above the plateau. This locality is well known for stone artifacts (mostly of quartzite) which have been determined by (excavations commenced in the 1930s) to be as old as 30,000 BP. The site also includes rock paintings. Small plutons of granite, e.g., by the Naabi Gate entrance into the Serengeti National Park (Fig. 7.1) are a subordinate component of the Mozambique Belt. The Mozambique Belt is also exposed in the Lake Eyasi Escarpment.

10.6 Ngorongoro Volcanic Complex

The Ngorongoro Volcanic complex is a group of eight, Pliocene age (3.7–1.79 Ma) volcanoes that cover an area of 90 km north-south and 80 km west-east. Five of the volcanoes reveal dome-shaped cones with gentle outer slopes, typical of shield volcanoes. Three volcanoes contain giant calderas. The internal walls of the calderas are steep and have suffered little erosion, enabling discrete ecosystems to be preserved within the centre of these near-circular features. The seven largest volcanoes are depicted on the satellite image (Fig. 10.1); one volcano, which has a small cone remains unnamed (Dawson 2008). The volcanism associated with the Ngorongoro Volcanic complex is dominated by lavas with abundant pyroclastics. Agglomerates occur near the summits of some cones, indicative of extinct vents. Some of the best exposures can be examined in the Engare Sero Gorge, which cuts the base of the escarpment near Lake Natron. The gorge reveals excellent exposures of thick beds of basaltic and nephelinite lavas and ashes, together with irregular debris flow deposits (Plates 5.2–5.4).

10.6.1 Ngorongoro

The Ngorongoro Volcano (2.5–1.9 Ma) has a base diameter of 35 km and contains one of the largest and best preserved calderas on Earth. The caldera shoulders attain a maximum height of 2,380 m, with the caldera floor located at an average elevation of 1,700 m (Figs. 5.7 and 10.3). Prior to caldera collapse, the cone is estimated to have attained a height of 5,000 m (Dawson 2008). The caldera is a near-circular feature, measuring 22 by 18 km, and with an internal area of approximately 370 km^2. The internal walls have an average height of 350 m. The caldera constitutes an internally-drained basin fed by two external streams. Several freshwater springs, including the scenic Ngoitokitok hot springs which supplies a small lake, support wetlands (Plate 10.1). The larger Lake Magadi is a shallow alkaline feature associated with deposits of sodium and magnesian carbonates. A detailed study of the clays has revealed that biotic CO_2 plays an important role in stabilizing the carbonates (Deocampo 2005). The extent and alkalinity of the lake varies considerably throughout the seasons. The size of Lake Magadi has varied considerably since the Late Pleistocene and carbon dating of ostracod-bearing clays has yielded dates for a high-stand at 24,000 BP. Parts of the caldera floor are covered by Pleistocene age lacustrine deposits (clays and silts) and the southwestern quadrant is covered by calcareous ash derived from Holocene eruptions of the Oldoinyo Lengai Volcano.

The average composition of the Ngorongoro Volcano is basaltic trachyandesite. Views of the southern caldera walls from the Ngoitokitok Springs enables the two main components of the volcano to be observed, basaltic lava flows (dark) and tephra-dominated sections (light). Flows of basalt and trachyte were erupted from ring fractures near the caldera walls. Two flows of rhyolite ignimbrite occur toward the top of the Lerai Section (the ascent road) which has been cut into the internal wall. They are associated with the final caldera event which is estimated to have occurred at approximately 2.0 Ma. The ignimbrite flows have yielded geochemical trends consistent with eruption of a stratified magma chamber with the silicic top and basaltic base inverted by sequential eruptions (Mollel et al. 2008). Lava flows also form low hills within the caldera as well as small scoria cones of basaltic composition, e.g., Engitati Hill. Lavas and ash from Ngorongoro occur within the Oldupai Gorge on the Eastern Serengeti Plains (McHenry et al. 2008).

10.6.2 Olmoti

The Olmoti Volcano ("cooking pot" in the Maasai language) has an overall diameter of 30 km. The central cone, which is visible to the north of the Ngorongoro Caldera (Plate 10.1) rises to a height of 3,101 m. The mountain is well known to the local inhabitants as it includes four small blow holes (diameters of a few tens of cm) that emit strong draughts of cold air. They are probably fossilised fumaroles (the volcano is extinct and there is no evidence of geothermal heat). Volcanic activity at Olmoti was very short-lived (2.01–1.79 Ma) and overlaps with the Ngorongoro volcanism, indicative of contemporaneous activity (Dawson 2008). Olmoti contains a large, yet shallow caldera within the centre of the cone. The caldera has a diameter of 6.5 km and depth of 100–200 m. The caldera flanks consist of thick sequences of lava. A small plug of trachytic lava and silicified breccias, approximately 130 m in height occurs within the caldera. Most of the caldera floor is covered by 50 m thick lacustrine deposits related to post-caldera collapse. The caldera includes a small lake. The southeastern flanks of the caldera have been breached by the Munge River which plunges by a series of waterfalls and a prominent ravine into the Ngorongoro Caldera (Fig. 10.3). The composition of Olmoti is basalt-trachyandesite in the lower part and trachyte in the upper part. Isotopic data suggests magmas were derived from melting of the lithospheric mantle, a conclusion also consistent with studies on the Ngorongoro volcanism (Mollel et al. 2009).

10.6.3 Empakaai

Empakaai, the most northerly of the volcanoes in the NCA is another giant feature with a diameter of approximately 30 km. The highest point of 3,235 m occurs on the Elanairobi Ridge which affords spectacular views northward into the rift valley, including of the Oldoinyo Lengai Volcano. The age of the Empakaai Volcano has not been determined and is assumed to be Pliocene. The volcano contains a thickly forested caldera with internal dimensions of 8 km by 6 km and a depth of almost 1,000 m (Plate 10.3b). Small parasitic cones occur on the floor of the caldera, which includes a deep, brackish lake with Pleistocene-Holocene sediments. The caldera walls consist

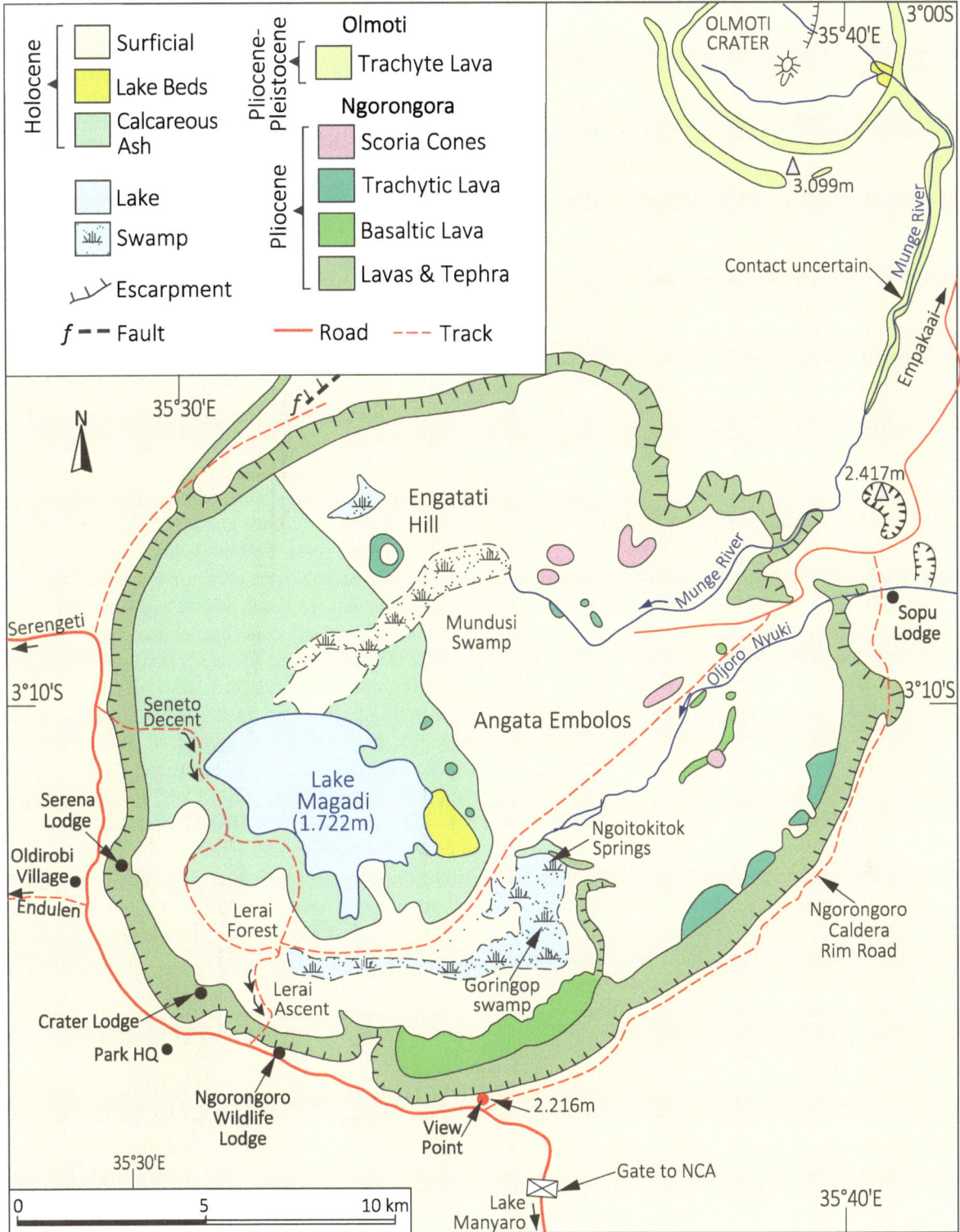

Fig. 10.3 Geological map of the Ngorongoro Caldera and part of the Olmoti Caldera, simplified from the Geological Survey of Tanzania 1:125,000 quarter degree sheet 53 (Pickering 1965)

of trachybasalts in the lower part and nephelinite-phonolite in the upper part.

10.6.4 Loolmalasin

The giant shield volcano of Loolmalasin, the highest peak in the Ngorongoro Highlands (3,648 m) has a diameter of 25 km. The southeastern and eastern slopes are steep, thickly forested, and include deep ravines close to the edge of the Western Escarpment. The volcano has not been dated but a Pliocene age is consistent with the basaltic and trachytic composition. There is no crater or caldera, although the rounded summit is disrupted by small faults. The Olosirwa Mountain (3,260 m) is considered to be part of the Loolmalasin Volcano. The unnamed volcano (2,624 m) to the east of Loolmalasin is comprised of basalt and nephelinite.

10.6.5 Sadiman, Lemagrut, and Oldeani

The three large volcanoes of Sadiman (also known as Satiman), Lemagrut (also known as Makarot), and Oldeani constitute the southwestern segment of the Ngorongoro Highlands. They are assumed to be of Pliocene age although only Sadiman has been radiometrically dated (3.7 Ma). Sadiman includes a thickly forested cone (2,860 m) dominated by tuffs and agglomerates with minor lava flows. The composition varies from nephelinite to phonolite (Zaitsev et al. 2012). Clasts of Sadiman lavas have been indentified at Oldupai Gorge (McHenry et al. (2008). The deeply incised cone of Lemagrut (3,130 m) towers some 1,300 m above the Laetoli site on the Eastern Serengeti Plains. Here, formation of tuffs and agglomerates was followed by minor flows of olivine basalt and trachybasalt. The agglomerate includes clasts of highly alkaline lava (nephelinite and jacupirangite) similar to those that have been washed into the Oldupai Side Gorge. The Oldeani cone (3,216 m) is a massive volcano with a small summit crater. Flows of basalt occur on the lower slopes but outcrop is poor and large parts are thickly forested. The western slopes of the Ngorongoro Highlands beneath the Lemagrut Volcano include the Ogol lavas (2.41 Ma), a sequence of olivine basalts erupted from parasitic cones on the Eastern Serengeti Plains.

10.7 Rift-Related Sedimentary Basins

The largest sedimentary basins in the NCA are associated with Lakes Eyasi and Manyara. They are dominated by Miocene-Holocene age sands and gravels eroded from the juxtaposed scarps (Dawson 2008). Basins also contain substantial volcaniclastic deposits. Details of the Oldupai and Laetoli basins can be found in Chap. 11.

10.8 Wind Blown Ash from Oldoinyo Lengai

The northern and western parts of the NCA are covered by thin deposits of ash derived from recent eruptions of the Oldoinyo Lengai Volcano (Chap. 17). The ash is dispersed by the prevailing easterly winds. The ash also covers the southwestern segment of the Ngorongoro Caldera, as noted above, which provides a localised area of nutrient-rich short grasses (similar to those on the Eastern Serengeti Plains) favoured by grazers in the dry seasons. The "Shifting Sands", a popular tourist destination located near Oldupai Gorge is a group of isolated black sand dunes (Fig. 10.2). They are unusual as most dunes occur in "fields" of yellow or red sands. The largest dune within the field has the classic crescent-shape of a barchan and is 9 m in height and 100 m in length (Plate 10.4). The individual dunes are migrating west at the remarkable rate of approximately 17 m/year. The tracks left behind are clearly visible and may include species of fossilised (extinct) dune beetles that have been widely studied. The sand was derived from Oldoinyo Lengai, probably from an eruption in 1940–1941. The composition is phonolite-nephelinite (associated with alkaline silicate magmas); they are not comprised of natrocarbonatite for which Oldoinyo Lengai is so renowned. The dunes consist almost entirely of coarse-grained ash; the finer ash has been removed by winnowing. The dispersion of the coarse-grained ash is suppressed by movement of the particles by wind, which together with bouncing of individual grains off the ground (the process of saltation), creates a negative charge. This electrical field may also assist with binding the grains together.

10.9 Treks

Treks can be arranged on some of the higher, forested slopes of the volcanoes, including the rim of the Ngorongoro Caldera and to the summit of the Olmoti Cone. The walk down the thickly forested slopes of the Empakaai Caldera to examine the brackish lake and salt deposits is recommended. A highlight of a visit to the Gol Mountains is a hike along the Ol Karien Gorge, a narrow slot in the quartzite only a few metres wide in some places. The valleys between the quartzite hills are dry and inhospitable in the dry season, but are lush and verdant after the rains when large herds of grazers spill over from the Serengeti Plains.

Plate 10.2 The Ngorongoro Caldera and the alkaline Lake Magadi in the wet (**a**) and dry (**b**) seasons

Plate 10.3 **a** The outer slopes of the Ngorongoro Volcano include nutrient-rich soils that are intensively farmed; **b** The Empakaai Caldera reveals thickly forested inner slopes and a central lake

Plate 10.4 **a** Isolated, barchan-style dune of black volcanic ash derived from Oldoinyo Lengai, Shifting Sands near Oldupai Gorge; **b** The front of the dune (which is migrating westward towards the right of the photograph) reveals a steep slope and distinctive outer horns

References

Dawson, J. B. (2008). The Gregory Rift Valley and Neogene-Recent Volcanoes of Northern Tanzania. *Geological Survey Memoir, 33,* 102 p.

Deocampo, D. M. (2005). Evaporative evolution of surface waters and the role of aqueous CO_2 in magnesian silicate precipitation: Lake Eyasi and Ngorongoro Crater, northern Tanzania. *South African Journal of Geology, 108,* 493–504.

McHenry, L. J., Mollel, G. F., & Swisher, C. C. (2008). Compositional and textural correlations between Olduvai Gorge Bed I tephra and volcanic sources in the Ngorongoro Volcanic Highlands, Tanzania. *Quaternary International, 178,* 306–319.

Mollel, G. F., Swisher, C. C., Feigenson, M. D., & Carr, M. J. (2008). Geochemical evolution of Ngorongoro Caldera, Northern Tanzania: implications for crust-magma interaction. *Earth and Planetary Science Letters, 271,* 337–347.

Mollel, G. F., Swisher, C. C., McHenry, L. J., Feigenson, M. D., & Carr, M. J. (2009). Petrogenesis of basalt-trachyte from Olmoti, Tanzania. *Journal of African Earth Sciences, 54,* 127–143.

Orridge, G. R. (1965). *Quarter Degree Sheet 69: Mbulu.* Mineral Resource Division of Tanzania.

Pickering, R. (1958). *Quarter Degree Sheet 38: Oldoinyo Ogol.* Geological Survey of Tanganyika.

Pickering, R. (1964). *Quarter Degree Sheet 52: Endulen.* Geological Survey of Tanzania.

Pickering, R. (1965). *Quarter Degree Sheet 53: Ngorongoro.* Geological Survey of Tanzania.

Zaitsev, A. N., Marks, M. A. W., Wenzel, T., Spratt, J., Sharygin, V. V., Strekopytov, S., et al. (2012). Mineralogy, geochemistry and petrology of phonolitic to nephelinitic Sadiman volcano, Crater Highlands, Tanzania. *Lithos, 152,* 66–83.

Oldupai Gorge and Laetoli

Abstract

The palaeoanthropological sites of Oldupai Gorge (previously known as Olduvai) and Laetoli are located on the Eastern Serengeti Plains, at the foot of the Ngorongoro Highlands, northern Tanzania. Hominin fossils and Oldowan stone tools have been unearthed from Plio-Pleistocene sequences of lacustrine sediments and volcanic ashes exposed in a shallow ravine carved by the ephemeral Oldupai River. The sediments and ashes accumulated in a small basin which formed due to warping of the regional plateau. Warping was triggered by the southward propagation of the East African Rift System (EARS). The presence of repetitive layers of water-lain volcanic ashes, or tuffs, derived from the Ngorongoro Volcanism, is particularly significant as they yield precise radiometric dates. Bed I of the Oldupai Group, where the first discoveries of *Zinjanthropus boisei* (OH-5) and *Homo habilis* (OH-7) were made by Mary Leakey, in 1959 and 1960, has an age of 2.015–1.803 Ma. The age of these two, partially intact hominin skulls has been determinded as 1.848 Ma (OH-5) and 1.848–1.832 (OH-7). Oldupai Gorge has also yielded fossils of *Homo erectus* (1.2–0.70 Ma) and the coexistence of different species of hominin is a unique feature of the locality. Reactivation of faulting associated with the Gregory Rift at approximately 1.15 Ma caused some of the older deposits to be reworked and buried beneath younger sediments and ashes. The youngest component of the Oldupai Group comprises wind-blown volcanic ashes from volcanoes located in the Gregory Rift Valley near Lake Natron. The Laetoli site has yielded hominin fossils and footprints from an area, where the Pliocene-age Laetoli Group is exposed in shallow gullies and areas of badland erosion. The most significant find is the 27-m-long trail made by *Australopithecus afarensis* (dated at 3.6 Ma), discovered by Paul Adell in 1978. The footprints are preserved in volcanic ash derived from the Sadiman Volcano.

Keywords

Badlands • Footprints • Hominin • Leakey Palaeoanthropology • Volcanism

Photographs not otherwise referenced are by the author.

© Springer International Publishing AG, part of Springer Nature 2018
R. N. Scoon, *Geology of National Parks of Central/Southern Kenya and Northern Tanzania*, https://doi.org/10.1007/978-3-319-73785-0_11

Plate 11.1 A view of the north side of Oldupai Gorge close to the visitor's centre reveals details of the lowermost formations of the Oldupai Group. Beds I and II (separated by the Marker Tuff) are dominated by pale-coloured sequences of clays, sands and volcanic ashes, with Bed III forming the well-defined buttes of red-coloured sands. The most significant hominin fossils have been excavated from Beds I and II. The Naabi Ignimbrite is exposed in the dry bed of the ephemeral river

11.1 Introduction

The world-famous palaeoanthropological sites of Oldupai Gorge and Laetoli occur within a section of the Ngorongoro Conservation Area (NCA) that extends onto the Eastern Serengeti Plains (Fig. 10.1). Both sites are dominated by areas of badland erosion, where ephemeral rivers have carved out shallow ravines. The discoveries made at Oldupai have greatly advanced our understanding of human evolution (e.g. Leakey 1984). The site has yielded the most continuous record of human evolution known, with more than 60 significant hominin discoveries having been made. The oldest finds at Oldupai have been unearthed from the lowermost sequences of the Oldupai Group, lacustrine sediments (deposited in a palaeo-lake) with successive layers of volcanic ashes and water-lain tuffs (Plate 11.1). Tuff layers are important as they are more resistant to erosion, enabling preservation of fossils, and can be radiometrically dated to obtain an accurate chronostratigraphy. The possibility of our human ancestors having evolved in Africa, particularly in areas proximal to the Ethiopian and Gregory Rifts is now widely accepted.

The spelling of Oldupai has recently been changed from 'Olduvai' to record the meaning as 'place of wild sisal' more accurately in the Maasai language. The Oldupai Visitors Centre has an excellent museum with replicas of most of the findings. Many of the original fossils are located in the National Museum of Tanzania, Dar-es-Salaam, and the British Museum, London. The first discoveries of hominin fossils made at Oldupai, particularly of *Zinjanthropus boisei* (OH-5) and *Homo habilis* (OH-7), made in 1959 and 1960, respectively, were widely reported. Evidence suggests *Homo sapiens* arrived comparatively recently at Oldupai, at approximately 17,000 BP, whereas southern African sites place the oldest known age at approximately 0.2 Ma, e.g. Pinnacle Point near Mossel Bay in the Southern Cape (Marean 2010). Laetoli occurs in a remote location to the west of the village of Endulen, approximately 25 km south of Oldupai. The visitor's centre here includes a life-size cast of the famous footprints made by *Australopithecus afarensis,* which were discovered by Paul Adell in 1978.

11.2 Location

The Eastern Serengeti Plains is an inhospitable environment that, for large parts of the year, is exceptionally arid. Most rivers flow westward into Lake Victoria, but the ephemeral Oldupai River drains eastward prior to petering out in the Olbalbal Swamps at the foot of the Ngorongoro Highlands (Fig. 10.2). The river has a total length of 48 km and is fed by two seasonal lakes, Lakes Masek and Ndutu. The Laetoli site occurs close to a watershed between two river systems, the Naisuri and Gadjingero Rivers, which flow northeast and northwest, respectively, into the Oldupai River, and the Garusi and Oldogom Rivers which drain southward into Lake Eyasi (Andrews and Bamford 2008). The description of the various river beds as gorges is rather misleading as incisions into the regional plateau are relatively shallow; Oldupai constitutes a steep-sided ravine (Plate 11.2). Most of the important discoveries at Oldupai have been made from a section close to the confluence with its main tributary, the Side Gorge (or Naisuri River), i.e. beneath the bluffs where the Visitor's Centre is located (Plate 11.3a, b). The footprints at Laetoli were discovered on the banks of a riverbed associated with an extensive area of badlands erosion (Plate 11.4a).

11.3 History of the Discoveries

Oldupai Gorge is synonymous with the husband and wife anthropological team of Mary and Louis Leakey, but it is less well known that these discoveries built on much earlier work undertaken in South Africa, principally by Raymond Dart and Robert Broom. Convincing a largely sceptical world that our human ancestors evolved in Africa, as related by Stringer and McKie (1996), has been a long process. Despite her initial lack of formal education, Mary Leakey stands out as one of the premier scientists in this field. Her finds include the first *Proconsul africanus* skull (discovered on Rusinga Island, Lake Victoria, Kenya in 1948), an extinct species of Great Ape that existed during the Miocene, as well as the discovery of *Zinjanthropus boisei* (OH-5) and *Homo habilis* (OH-7) at Oldupai. Louis Leakey is well known for his contributions to both hominin research and the study of primates.

Louis Leakey's interest in hominins was in part inspired by his excavations of sites in northern Kenya (particularly around Lake Turkana), where he had previously collected both hominin fossils and stone tools, an area where Richard Leakey has subsequently made many important discoveries. The initial finds by the Leakey's at Oldupai were restricted to stone tools, known as Oldowan for the importance of this type locality, but it was the discovery of the cranium OH-5, also known as 'Nutcracker Man', in recognition of the powerful jaws and teeth, that proved so significant. This subgroup of *Australopithecine* is also known as *Paranthropus*; although many paleoanthropologists have reverted to the original name of *Zinjanthropus* (the name *boisei* acknowledges the benefactor of the original research, Charles Boise). The footprints of *Australopithecus afarensis* discovered at Laetoli by Paul Adell were first described and interpreted by Leakey (1981). The type locality for this genus (including the famous 'Lucy' skeleton dated at 3.2 Ma) is, however, in Ethiopia (e.g. Stringer and McKie 1996).

The first scientist to have visited Oldupai was probably Wilhelm Kattwinkel, a German entomologist, who in 1911 discovered various animal fossils, including an extinct three-toed horse. The first hominin remains to be discovered at Oldupai, however, were found in 1913 by Hans Reck, a German palaeontologist and volcanologist (Reck also submitted the first detailed report on the Oldoinyo Lengai Volcano). The fossils discovered by Hans Reck at Oldupai, known as the 'Reck Skeleton', and subsequently categorised as *Homo erectus*, are surrounded by mystery as they disappeared from the Munich Museum where they had been housed during WWII. Hans Reck faced a considerable problem with getting his early work accepted as there were no means of dating the rocks prior to the advent of radiometric techniques in the late 1950s. His estimation of the age of *Homo erectus* as 0.50 Ma was dismissed at the time as far too old.

The Leakey's interest in Oldupai was stimulated by visits in 1927 and 1929 to examine the Reck Skeleton in the Munich Museum and to meet Hans Reck in the Natural History Museum in Berlin. After examining rock samples from Oldupai, Louis Leakey became convinced the geology was similar to sites in northern Kenya. He also intimated some of the rock samples were stone tools. Despite differences of opinions on the stone tools and the age of the Reck Skeleton (estimated by Leakey at 20,000 BP), Reck was invited to join the first of the Leakey's expeditions. On arrival at Oldupai Gorge in September 1931 the team immediately discovered numerous hand axes made from lava (not flint as in European sites). The field evidence caused Louis Leakey to agree with the much older age of the skeleton preferred by Hans Reck (Leakey et al. 1931).

11.4　Geology of Oldupai

The Oldupai Gorge cuts eastward from the Central African Craton (in the vicinity of Lakes Masek and Ndutu), initially onto quartzite of the Mozambique Belt, and then onto the Ngorongoro Volcanics (Fig. 10.2). The quartzite is exposed where the main road from the Naabi Gate in the Serengeti crosses the river, as well as at Naibor Hill on the north side of the gorge near the visitor's centre. The Naabi Ignimbrite, a well-defined layer of black and lustrous pyroclastic rock, which has a thickness of approximately 7 m is the principal capping of the Ngorongoro Volcanics in this area (Plate 11.2a). The ignimbrite, which is linked to the catastrophic caldera event of the Ngorongoro Volcano (McHenry et al. 2008) is unconformably overlain by the Oldupai Group. The age of the ignimbrite has been determined very accurately as 2.038 ± 0.005 Ma (Deino 2012); this places a lower age limit on potential discoveries at Oldupai (Fig. 11.1a).

The geology of the Oldupai Group has been described by American geologist Richard Hay, who studied the area in great detail over a period of more than 40 years (Hay 1976). The bulk of these rocks occur beneath the regional plateau (they are obscured by a thin, recent covering of volcanic ash derived from the Oldoinyo Lengai Volcano) and exposures are restricted to a linear section in the gorge. This has revealed a marginal section of the palaeo-basin. The maximum thickness of the Oldupai Group is generally estimated as 100 m, but a recent drilling programme suggests that this is increased considerably.

The lowermost formations of the Oldupai Group were deposited in a palaeo-lake which formed during the Late Pliocene and Early Pleistocene (2.0–1.15 Ma) due to localised warping of the regional plateau (Hay 1976). Warping was triggered by rifting which propagated southward from Kenya into northern Tanzania near Lake Natron. Without the EARS, it is unlikely that sites suitable for preservation of the hominin fossils would have developed. Reactivation of significant faulting at approximately 1.15 Ma had the effect of reworking and exposing the lowermost formations. The palaeo-lake dried entirely during the Late Pleistocene at approximately 0.5 Ma.

11.5　Details of the Oldupai Group

Hay (1976) retained much of the stratigraphy originally developed by Reck during his visit in 1913, recognising seven formations, or beds (Fig. 11.2a). The lowermost beds (I–II) are dominated by lacustrine sediments (clays and sands). They are intercalated with thin layers of volanic ash and tuffs derived from the waning stages of the Ngorongoro volcanism i.e. after formation of the Naabi Ignimbrite. These beds have yielded the most important fossil discoveries. The intermediate beds (III–IV) are dominated by sediments, with the latter containing numerous fossils. The uppermost beds (V–VII) are mostly comprised of ash derived from relatively younger eruptions of volcanoes located near Lake Natron. The Oldupai Group reveals a number of clear depositional breaks (unconformities), both between and in some instances within formations (Hay 1976; Blumenschine et al. 2012).

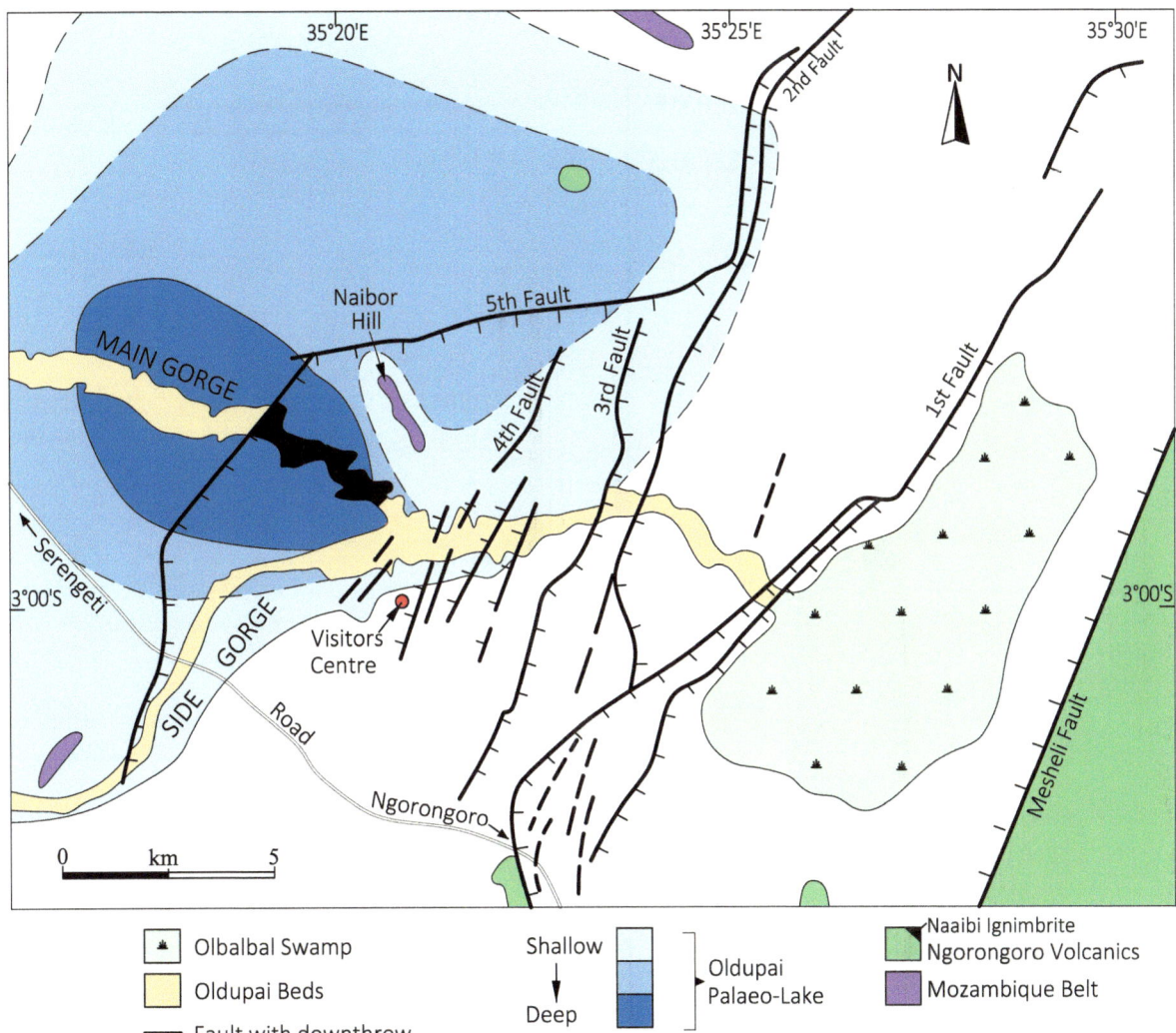

Fig. 11.1 Geological map of the Oldupai Basin based on Stollhofen and Stanistreet (2012), showing the estimated extent of the palaeo-lake (original modified from earlier studies including Hay 1994). In the Late Pleistocene, localised tectonism of the regional plateau west of the rift valley caused the palaeo-lake to start draining along the eastward-flowing Oldupai River system. The river eroded some of the earlier-formed (Pliocene and Early Pleistocene) beds within the Oldupai Group that had originally been deposited unconformably on the Ngorongoro Volcanics

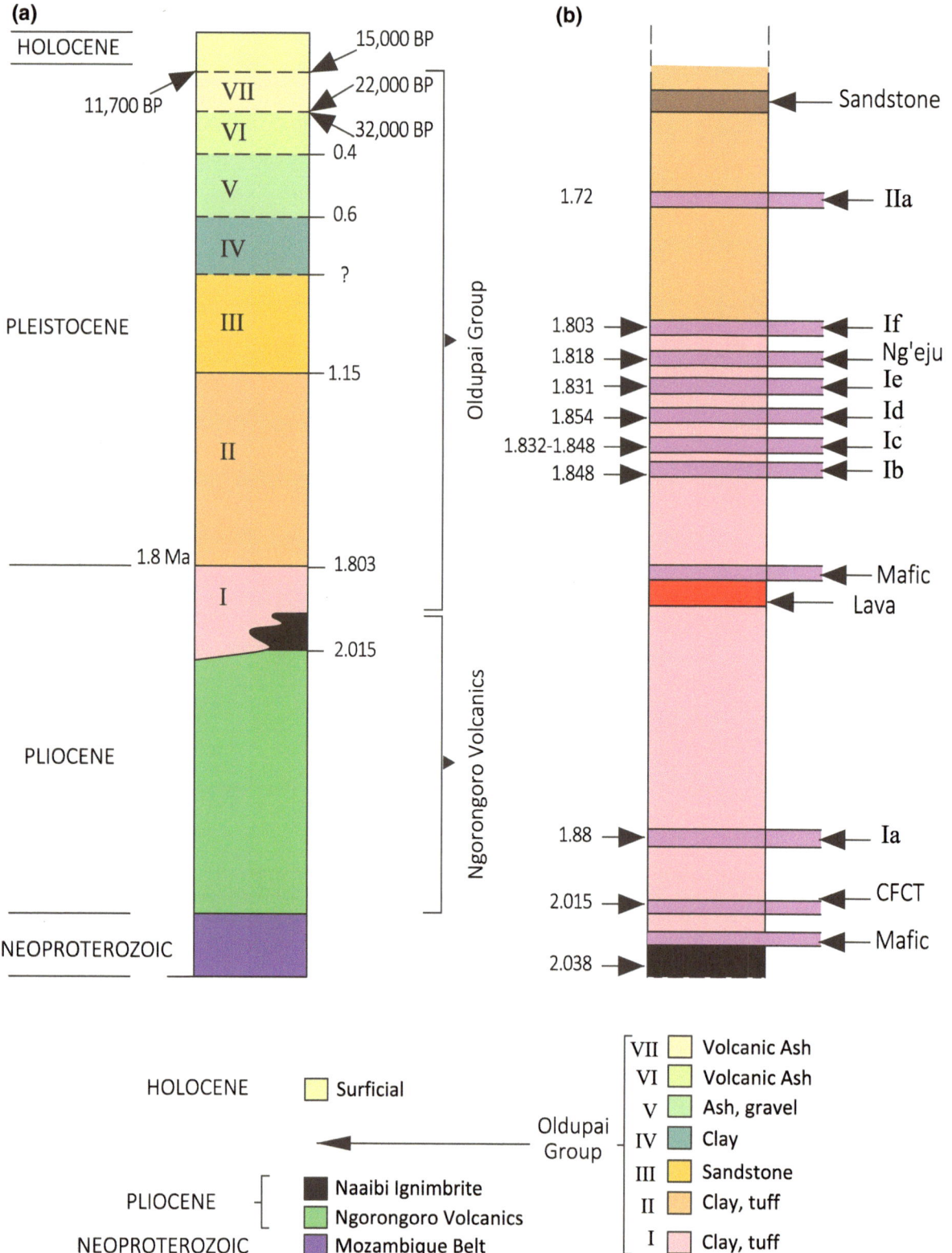

Fig. 11.2 **a** Stratigraphy of Oldupai Gorge with estimated ages from various sources including Hay (1994), Deino (2012), and Blumenschine et al. (2012); **b** Details of the Pliocene-age Bed I and the lowermost part of the Pleistocene-age Bed II modified from Stollhofen and Stanistreet (2012), but with radiometric dates of Deino (2012)

11.5.1 Lowermost Beds I–II

Bed I is a 50-m-thick sequence dominated by pale-coloured clays and volcanic tephra. The Late Pliocene age is constrained by radiometric dating of the volcanic components between 2.015 and 1.803 Ma (Deino 2012). A number of marker layers are recognised, including hard, compact layers of water-lain tuffs (labelled Ia through If). The tuff layers have been dated extremely accurately (Fig. 11.2b). Dates used here are from Deino (2012), but there are other dates that are not entirely consistent with these data (e.g. McHenry 2012). Most of the tuff layers represent unconformities. Tuffs Ia through Ie are related to major eruptions of Ngorongoro that occurred after the caldera event (McHenry et al. 2008). Tuff Ie is a particularly important marker as it is laterally very persistent. Bed I also contains eroded lava clasts associated with the Sadiman Volcano (3.7 Ma) that were transported into the palaeo-lake. Bed I hosts two of the most significant discoveries made at Oldupai, the initial discovery of *Zinjanthropus boisei* (Plate 11.3a) as well as that of *Homo habilis* (Plate 11.3b).

Bed II is a 35-m-thick succession of alternating sequences of clay and sandstone with minor tephra. It spans the transition from the Pliocene to the Lower Pleistocene. It is separated from Bed I by Tuff If, also known as the Marker Tuff which has been dated at 1.803 Ma (Fig. 11.2b). Tuff If is an important transition: the source of the tephra switched from Ngorongoro to the Olmoti Volcano (McHenry et al. 2008). The morphology of Bed II was affected at approximately 1.5 Ma when the lacustrine environment of the western part of the basin changed to grassland. The upper age limit is constrained at 1.15 Ma. Bed II is the height at which Reck found his skeleton of *Homo erectus*.

11.5.2 Intermediate Beds III–IV

The ages of the intermediate formations, Beds III and IV, are poorly constrained at between 1.15 and 0.6 Ma (Hay 1994). Bed III is a 10-m-thick succession of reddish-coloured sandstones with minor conglomerates and clays. This formation can be observed in the prominent buttes that cap the sidewalls of the ravine (Plate 11.2b). The reddish colour is correlated with periodic changes from wet to dry along the fringes of small alluvial fans. Few fossils have been discovered at this height. The thickness and colouration of Bed III is very variable and the reddish colour is not present in the west. There is a significant unconformity between Beds III and IV. The latter has an average thickness of approximately 6 m in the eastern part of the gorge, but thickens considerably towards the west. Bed IV comprises fluvial sediments in the lower part (clay, conglomerate), with aeolian sands in the upper part. Bed IV has yielded abundant fossils, including of *Homo erectus* and provided evidence that the latter hunted a profusion of different game species. Several land surfaces have been identified in Beds III and IV which are correlated with the rapid extinction of large numbers of wildlife species during arid periods.

11.5.3 Uppermost Beds V–VII

The 25-m-thick Masek Bed (V) consists of river gravels and volcanic ash. The age is estimated at 0.6–0.4 Ma (Hay 1994). The Ndutu (VI) and Naishiusiu (VII) Beds are comprised of volcanic ash, with ages estimated at 0.4 Ma-32,000 BP and 22,000-15,000 BP, respectively. Beds V and VI contain few fossils. Bed VII provides the oldest fossil evidence of *Homo sapiens* in the area. The volcanic ashes of Beds V through VII were derived from the alkali basalt and natrocarbonatite volcanoes located near Lake Natron, with the source switching from Kerimasi (Bed V) to Oldoinyo Lengai (Beds VI–VII) over time (Fig. 17.1).

11.5.4 Faulting

Oldupai Gorge is cut by a sequence of faults that traditionally have been numbered from east to west (Fig. 11.1). Detailed investigations led Stollhofen and Stanistreet (2012) to recognise discrete fault-bound compartments that greatly influenced sedimentation patterns. The significant faulting at 1.15 Ma is linked with the rejuvenation of the fault associated with the Western Escarpment, in the vicinity of Lake Natron. This caused the palaeo-lake at Oldupai to shrink considerably. The lowermost beds were subsequently reworked by fluvial activity associated with the deposition of the intermediate beds. At approximately 0.5 Ma, localised tectonism caused the lake to drain entirely along an eastward-flowing river that eroded and reworked both the lowermost and intermediate beds. The uppermost beds were deposited on new land surfaces after this phase of tectonism.

11.6 Geology of Laetoli

The Laetoli Beds are divided into a Lower Unit (4.3–3.8 Ma) that contains few fossils and an Upper Unit (3.8–3.5 Ma) that is rich in both fossils and tracks (Hay 1987; Drake and Curtis 1987). The tephra that contains the famous footprints is one of 18 layers with a combined thickness of 15 cm deposited over an estimated period of a few weeks and dated at 3.6 Ma. An earlier hypothesis in which the Laetoli Beds were related to the Lemagrut Volcano (Leakey and Hay 1979) may be incorrect, as recent evidence suggests the source is the Sadiman Volcano (Zaitsev et al. 2011). The Laetoli Beds are overlain by the Ndulanya Beds (3.5–2.4 Ma), the Ogol Lavas (2.4 Ma), the Naibadad Beds (2 Ma), the Olpiro Beds (1.2 Ma) and the Ngaloba Beds (0.15 Ma–30,000 BP).

The Laetoli area is subjected to badland erosion, with new gullies in the river beds persistently being opened (Plate 11.4b). In 1979, the hominin footprints were buried under a rock pavement for protection (Plate 11.4c). Casts of the tracks can be observed in the Laetoli Museum (Plate 11.4d). Masao et al. (2016) have recently discovered a new track made by two individuals approximately 150 m from the original site, travelling in a similar direction as the early footprints. This supports the theory that the body size of *Australopithecus afarensis* varied considerably amongst individuals. Additional fossils and tracks can be predicted to be unearthed due to the extensiveness of the badland erosion.

11.7 Hominins

The discovery by Mary Leakey of OH-5, the cranium of an *australopithecine* that Louis Leakey identified and named *Zinjanthropus boisei* was located at level 22 in a trench at the FLK site (named for Louis's first wife, Frida Leakey Korongo). The FLK-22 or FLK-*zinj* level is a transgressive surface located near the base of Tuff Ic. This surface transgresses down through the clays and sands towards Tuff Ib. The lower jaw and parts of a hand of OH-7, *Homo habilis*, were discovered at FLK-South, a few hundred metres from the original trench, at a similar height. Tuff 1b is a persistent marker in this area, but Tuff Ic rests on increasingly lower levels to the north of the FLK site. Rapid deposition of Tuff Ic enabled the underlying clays and sands, together with their contained hominin fossils to be almost perfectly preserved. Two volcanic fallout events are recognised between Tuff Ib and Ic. The lowermost component is a 1–2-cm-thick layer of ash with pumice. The uppermost is a size-graded unit of coarse-to-fine ash, typical of subaqueous volcanic deposits. The OH-5 and OH-7 fossils were originally dated at 1.75 Ma, but data of Deino (2012) suggests ages of between 1.848 and 1.832 Ma based on their location between Tuffs 1b and Ic, respectively (Fig. 11.2b).

A cast of the cranium of *Zinjanthropus boisei* (OH-5), discovered in Bed I at Oldupai Gorge together with the Peninj mandible, the latter being found by Kamoya Kimeu in Peninj, near Lake Natron is shown in Plate 11.5a. Other notable finds at Oldupai include OH-8 (foot of *H. habilis*) and OH-9 (flattened cranium of *H. habilis* also known as 'Twiggy'). Additional discoveries of *Homo habilis*, including a complete skull have been made in other localities in East Africa, most notably at Koobi Fora on the eastern shores of Lake Turkana in Kenya (Plate 11.5b).

Blumenschine et al. (2003, 2012) present detailed studies (as well as useful overviews of the geology) in which the environment of Beds I and II has been reinterpreted as a wooded peninsula of a lake which hominins visited (and hunted in), rather than settled semi-permanently as was previously envisaged. These interpretations are supported by the recent discovery of a fossilised tree in Bed I at Oldupai, described in a comprehensive article and overview by Habermann et al. (2016).

Palaeoanthropological evidence suggests *Zinjanthropus boisei* occurred in East Africa during the period between 2.3 and 1.2 Ma. For part of this period, notably at Oldupai Gorge, they coexisted with *Homo habilis*, the latter having occurred in this area between 1.8 and 1.2 Ma. The *Zinjanthropus* has distinctive, robust features in comparison to the more modern *Homo habilis*, although some researchers believe that *Homo habilis* should be reclassified with the more primitive *Australopithecus*. *Homo habilis* differs from the *Australopithecine* or *Zinjanthropus* on the basis of larger brain capacity and decreased teeth size. The fossils of *Homo erectus* discovered at Oldupai cover the period between 1.2 and 0.7 Ma. categorisation of hominin fossils is contentious and the story of how the discoveries at Oldupai were received into the scientific community is full of intrigue as related by Meredith (2011).

The evolutionary relationship between these species, and particularly to the one that concerns us most (*Homo sapiens*) is unlikely to be ever fully proven (due to the paucity of evidence), but traditional argument that evolution occurred in a regular, tree-like pattern with each subsequent jump being more evolved is probably best discarded. The evolutionary patterns are far more complex and in the words of geneticist Adam Rutherford 'there was no beginning, and there are no missing links, just the ebb and flow and ebb again of living through epochs' (Rutherford 2017).

11.8 Other Fossils and Oldowan Tools

Oldupai has yielded numerous fossils of mammal species that have gone extinct, indicative of relatively rapid extinction patterns accompanying the diverse speciation that characterises East Africa. At Laetoli, footprints of large antelopes, elephant, and horses, as well as raindrops, have been discovered suggesting the hominins were part of Pliocene-age migrations during rainy seasons. Evidence of increased development of hominins at Oldupai is supported by studies of Oldowan stone tools. The Leakey's excavated more than two thousand tools which cover the period between 1.85 and 0.6 Ma. Mary Leakey observed that *Homo habilis* used several different types of stone tools and achieved a high level of social development, including a stone circle (for anchoring wooden huts) comparable to those used today by tribal groups in northern Kenya. The oldest Oldowan tools from Bed I consist of pebbles with edges that have been sharpened. The overlying beds reveal more advanced tools including hand axes. Some of the tools were carved from lava associated with the Sadiman Volcano, dense homogeneous rocks with an attractive greenish colour.

Plate 11.2 a View looking towards the Visitor's Centre at Oldupai Gorge close to the confluence with the Side Gorge shows how the ephemeral river has eroded down onto the resistant Naabi Ignimbrite (black), with Bed I (pale grey), Bed II (grey) and Bed III (reddish-brown) restricted to the sidewalls; **b** View looking east over the buttes of Bed III towards the Ngorongoro Highlands from where the tephra in Beds I and II was sourced

Plate 11.3 a The discovery site of *Zinjanthropus boisei* (FLK) in Bed I, which occurs beneath a sequence of clays (light) and volcanic ashes (dark) is marked by a stone cairn and plaque, Oldupai Gorge; **b** The lower jaw and parts of a hand of *Homo habilis* were discovered at the FLK-South site, also in Bed I, Oldupai Gorge

Plate 11.4 **a** Laetoli is located in an area of badland erosion; the museum is visible in background; **b** New gullies continue to develop in the river bed at Laetoli; **c** View of the rock pavement under which the footprints at Laetoli are buried, looking northeast towards the riverbed with extinct volcanoes of the Ngorongoro Highlands in the background; **d** Reconstruction and cast of the footprints made by *Australopithecus afarensis*, Laetoli Museum

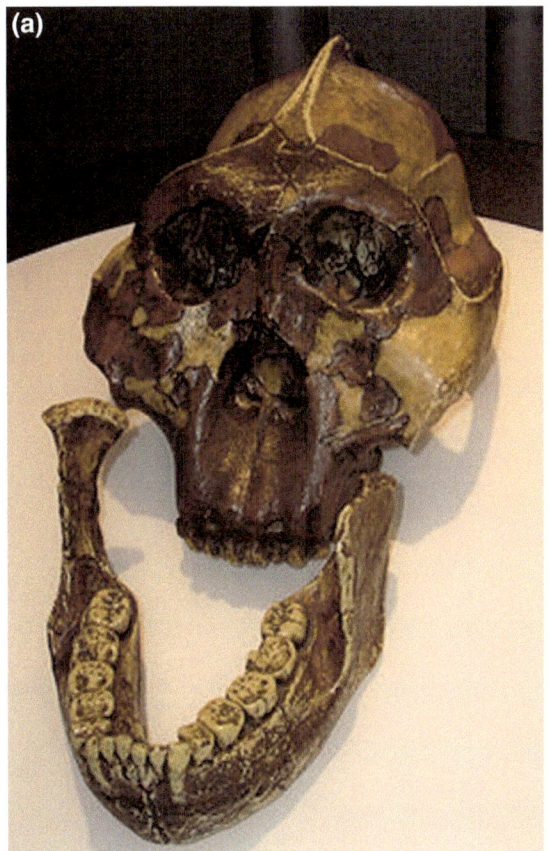
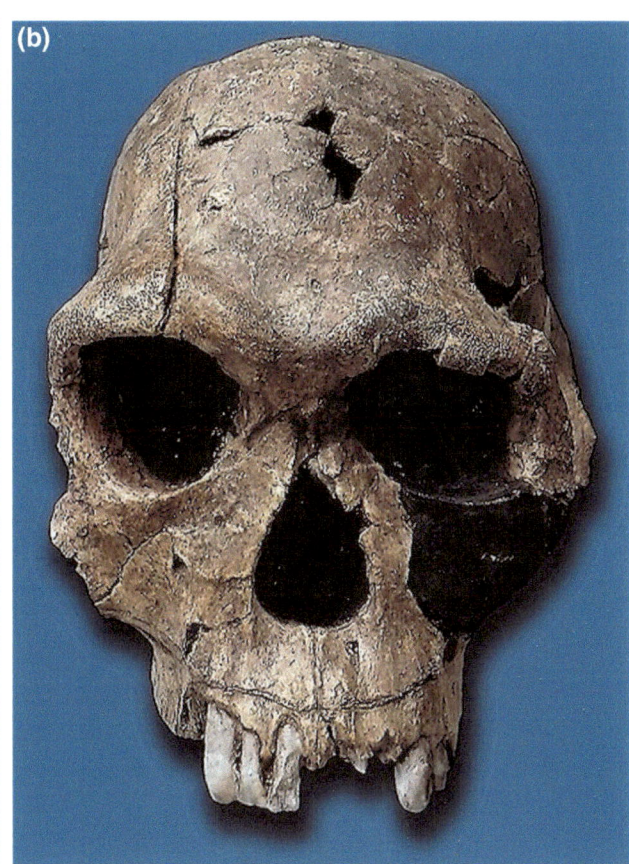

Plate 11.5 **a** Reconstruction of the cast of the 1.85 Ma-old cranium of *Zinjanthropus boisei* (OH-5) discovered in Bed I at Oldupai Gorge by Mary Leakey in 1959, with the lower jaw (Peninj mandible) found by Kamoya Kimeu in Peninj, near Lake Natron, in 1964. Photograph by Lillyundfreya of the Westfälisches Musuem; **b** Replica of the 1.9 Ma-old cranium of *Homo habilis* (KNM-ER1813) which was discovered by Kamoya Kimeu at Koobi Fora, Kenya, in 1973. Photograph by Locutus Borg

References

Andrews, P., & Bamford, M. (2008). Past and present vegetation ecology of Laetoli, Tanzania. *Journal of Human Evolution, 54,* 78–98.

Blumenschine, R. J., Peters, C. R., Masao, F. S., Clarke, R. J., Deino, A. L., Hay, R. L., et al. (2003). Late Pliocene Homo and hominid land use from western Olduvai Gorge, Tanzania. *Science, 299,* 1217–1221.

Blumenschine, R. J., Stanistreet, I. G., Njau, J. K., Bamford, M. D., Masao, F. S., Albert, R. M., et al. (2012). Environments and hominin activities across the FLK peninsula during Zinjanthropus times (1.84 Ma), Olduvai Gorge, Tanzania. *Journal of Human Evolution, 63,* 364–383.

Deino, A. L. (2012). ^{40}Ar/^{39}Ar dating of Bed I Olduvai Gorge, Tanzania and the chronology of early Pleistocene climate change. In R. J. Blumenschine, E. T. Masao, I. G. Stanistreet, & C. C. Swisher (Eds.). Five decades after Zinjanthropus and Homo habilis: Landscape paleoanthropology of Plio-Pleistocene Olduvai Gorge, Tanzania. *Journal of Human Evolution 63,* 252–273.

Drake, R., & Curtis, G. H. (1987). K-Ar chronology of the Laetoli fossil localities. In M. D. Leakey & J. M. Harris (Eds.), *Laetoli, a Pliocene Site in Northern Tanzania* (pp. 48–52). Oxford: Clarendon Press.

Habermann, J. M., Stanistreet, I. G., Stollhofen, H., Albert, R. M., Bamford, M. K., Pante, M. C., et al. (2016). In situ 2.0 Ma trees discovered as fossil rooted stumps, lowermost Bed I, Olduvai Gorge, Tanzania. *Journal of Human Evolution, 90,* 74–97.

Hay, R. L. (1976). *Geology of the Olduvai Gorge: A study of sedimentation in a semiarid basin* (203 p). Berkeley: University of California Press.

Hay, R. L. (1987). Geology of the Laetoli Beds. In M. D. Leakey & J. M. Harris (Eds.), *Laetoli, a Pliocene Site in Northern Tanzania* (pp. 23–47). Oxford: Clarendon Press.

Hay, R. L. (1994) Geology and dating of Beds III, IV and the Masek Beds. In: M. D. Leakey & D. A. Roe (Eds.), Olduvai Gorge volume 5: Excavations in Beds III, IV and the Masek Beds, 1968–1971 (pp. 8–14) Cambridge: Cambridge University Press.

Leakey, M. D. (1981). Discoveries at Laetoli in Northern Tanzania. *Proceedings of the Geologists' Association, 92*(2), 81–86.

Leakey, M. D. (1984). *Disclosing the past* (224 p). New York: Doubleday and Co.

Leakey, M. D., & Hay, R. L. (1979). Pliocene footprints in the Laetoli Beds at Laetoli, Northern Tanzania. *Nature, 278,* 317–323.

Leakey, L. S. B., Hopwood, A. T., & Reck, H. (1931). Age of the Oldowan bone beds, Tanganyika. *Nature, 128,* 724.

Marean, C. (2010). Pinnacle point cave 13B (Western Cape Province, South Africa) in context: The Cape Floral Kingdom, shellfish, and modern human origins. *Journal of Human Evolution, 59,* 425–443.

Masao, F. T., Ichumbaki, B. T., Cherin, M., Barili, A., Boschian, G., Lurino, D. A., Menconero, S., Moggi-Cecchi, J., & Manzi, G. (2016). New footprints from Laetoli (Tanzania) provide evidence for marked body size variation in early hominins. *eLife, 5,* 1–29.

McHenry, L. J. (2012). A revised stratigraphic framework for Olduvai Gorge Bed I based on tuff geochemistry. In R. J. Blumenschine., E. T. Masao, I. G. Stanistreet & C. C. Swisher (Eds.), Five decades after *Zinjanthropus* and *Homo habilis*: Landscape Paleoanthropology of Plio-Pleistocene Olduvai Gorge, Tanzania. *Journal of Human Evolution, 63,* 284–299.

McHenry, L. J., Mollel, G. F., & Swisher, C. C. (2008). Compositional and textural correlations between Olduvai Gorge Bed I tephra and volcanic sources in the Ngorongoro Volcanic Highlands, Tanzania. *Quaternary International, 178,* 306–319.

Meredith, M. (2011). *Born in Africa: The quest for the origins of human life* (230 p). London: Simon and Schuster.

Rutherford, A. (2017). *A brief history of everyone who ever lived* (419 p). London: Weidenfeld and Nicolson.

Stollhofen, H., & Stanistreet, I. G. (2012). Plio-Pleistocene synsedimentary fault compartments, foundation for the eastern Olduvai Basin paleoenvironment, Tanzania. *Journal of Human Evolution, 63,* 309–327.

Stringer, C., & McKie, R. (1996). *African exodus* (267 p). London: Jonathan Cape.

Zaitsev, A. N., Wenzel, T., Spratt, J., Williams, T. C., Strekopytov, S., Sharygin, V. V., et al. (2011). Was Sadiman Volcano a source for the Laetoli Footprint Tuff? *Journal of Human Evolution, 61*(1), 121–124.

Abstract

The free-standing mountain of Kilimanjaro is a giant edifice comprised of three discrete volcanoes. They collectively cover an area of 6,000 km^2. The two lowest peaks, Shira and Mawenzi are eroded remnants of extinct cones. The highest peak, Kibo is an eroded caldera with an extensive summit plateau that includes the Reusch Crater. Shira is part of the Older Volcanism of northern Tanzania (2.5–1.9 Ma), but Mawenzi (1.0–0.45 Ma) and Kibo (0.48–0.15 Ma) are associated with the Younger Volcanism. Kibo displays a differentiation trend in which trachyandesite gives way initially to phonolite and then to nephelinite, characteristic of the explosive volcanism associated with many volcanoes in the Gregory Rift. The mountain reveals successive botanical zones, including montane forest, heath and moorland, Alpine desert and icefields. There is evidence of multiple cycles of ice advance and retreat during the Late Pleistocene Ice Ages. Cores drilled into the Northern Ice Field, Kibo, provide information on the age of the ice and climatic cycles during the Holocene epoch. The rapidly receding icefields and glaciers are relics of ice that has a maximum age of 11,700 BP. Current trends suggest the ice may disappear in the near future. The Reusch Crater reveals evidence of geothermal heat and Kibo is categorised as a dormant volcano although there is no definite evidence of historical activity. Large parts of the mountain are protected in a national park that attracts huge numbers of visitors. There are seven main trekking routes offering the opportunity to reach Uhuru Peak on Kibo (5,895 m) with little technical difficulty. The ascent of Mawenzi, a precipitous cone which is the remnant of an eroded crater with near-vertical outer cliffs and steep ravines on the outer slopes is rarely attempted.

Keywords

Botanical zones • Caldera • Ice recession
Kibo • Pleistocene glaciation • Reusch Crater
Volcanism

Photographs not otherwise referenced are by the author.

© Springer International Publishing AG, part of Springer Nature 2018
R. N. Scoon, *Geology of National Parks of Central/Southern Kenya and Northern Tanzania*, https://doi.org/10.1007/978-3-319-73785-0_12

Plate 12.1 Western slopes of Kibo viewed from Mount Meru. Uhuru Peak occurs on the caldera rim near the Southern Ice Field (right), with the relicts of the Northern Ice Field just visible on the ridge (left)

12.1   Introduction

Kilimanjaro is one of the world's largest free-standing mountains and the highest peak of Kibo towers more than 5,000 m above the regional plateau (Plate 12.1). Kilimanjaro is comprised of three discrete volcanic centres, Shira, Mawenzi and Kibo, which cover an area of 80 by 48 km. The two lowest peaks, Shira (4,006 m) and Mawenzi (5,149 m) are deeply eroded. The massif is located 80 km east of the Gregory Rift in the structurally complex northern Tanzania divergence, an area characterised by numerous giant volcanoes (Fig. 5.4). The annular vegetation zones which characterise the larger East African volcanoes are particularly well developed on Kilimanjaro. The original reports in 1848 by Johannes Rebmann of an ice-capped peak near the Equator were treated with scepticism in Europe, despite references to an African 'Moon Mountain' (possibly the Ruwenzori) in classical literature, i.e. by the Greek geographer Strabo and the Roman mathematician Ptolemy. The first ascents of Kibo were made by Hans Meyer, who reached the summit plateau via the Southern Ice Field in 1887 and Uhuru Peak in 1889. Most of the volcanic edifice is incorporated into the Kilimanjaro National Park, which is also a World Heritage Site. The park is a major contributor to foreign earnings and is visited by large numbers of trekkers. The mountain is usually approached from the southern side via the regional town of

Moshi (altitude of 854 m), but views from the northern (Kenya) side are equally spectacular (Plate 12.2a).

12.2   Botanical Zones

The successive botanical and climatic zones on Kilimanjaro are delineated by the altitude (Mitchell 1971; Williams et al. 1967). The lower slopes (1,900–3,000 m) are covered by dense montane forests, although some sections have been replaced by commercial farms and shambas (Fig. 12.1). The southern and western slopes are most favourable for cultivation as they receive the highest rainfall and have more nutrient-rich soils. The wetter forests are dominated by camphor, podocarpus and fig trees, with junipers and olive trees occurring in drier sections. Vines, Old Mans Beard and ferns are important components. The forest is replaced upwards, initially by a belt of bamboo, and then by a zone of heather and moorland with a maximum range in elevation of 2,750–4,300 m. The upper moorland (above 3,500 m) includes species of giant groundsels and lobelia that are unique to East Africa. Alpine desert (a zone with very little vegetation), or high-altitude tundra prevails above 4,300 m and is notably well developed in the Saddle, i.e. between Kibo and Mawenzi (Plate 12.2b). The Alpine zone was severely glaciated during the Late Pleistocene. The current lower limit of the Holocene slope glaciers is approximately

4,900 m. Above this height, flowering plants cannot survive and are replaced by lichen and moss.

12.3 Regional Geology

The Kilimanjaro edifice occurs in a 100-km-long faulted belt of intense volcanism that strikes almost at right angles to the Gregory Rift (Dawson 2008). The location corresponds with a prominent NNW striking fault that intersects a pronounced structural lineament, aligned at 80°, associated with other volcanoes in the area, including Meru, Tarosera and Essimingor (Fig. 5.7), as described by Nonnotte et al. (2008). The three volcanic centres that constitute Kilimanjaro are constructed on older volcanic terranes within the divergence, including the Pliocene-age Molog basalts to the northwest of Kibo (Fig. 12.2). Mawenzi and Shira, however, may in part overlap onto the Mozambique Belt.

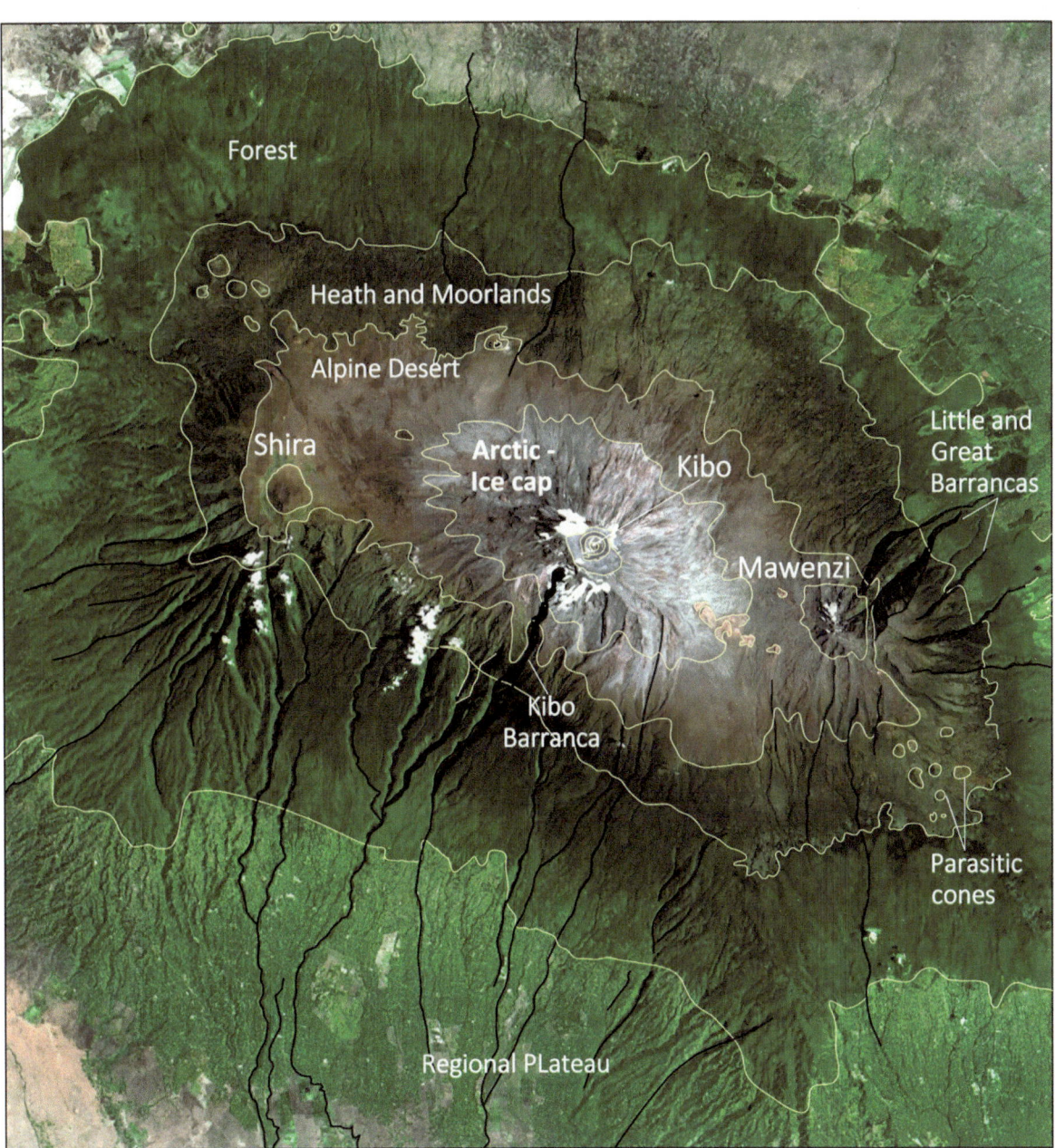

Fig. 12.1 Image of the Kilimanjaro massif showing part of the National Park and the successive botanical zones. The lower slopes include large areas, where the montane forest has been cleared for farmland. The Reusch Crater on the summit of Kibo is clearly visible. *Source* Google-Earth Image 2017, NASA, Image US Geological Survey

12.4   Volcanic Centres

Kilimanjaro was mapped in remarkable detail during the 1953 and 1957 joint-expeditions of the University of Sheffield and the Geological Survey of Tanganyika. A report on the first expedition (Wilcockson 1956) was followed by a comprehensive treatise (Downie and Wilkinson 1972). The following is largely summarised from these accounts together with contributions of Jennings (1971) and Wilkinson et al. (1983). The map presented here (Fig. 12.2) is simplified from the original mapping which was undertaken at a scale of 1:125,000 and is attached with the 1972 report. Volcanism associated with the composite massif covers an area of 6,000 km^2, considerably larger than the topographic expression as lavas extend onto the adjacent plateau.

The three volcanic centres of Shira (west), Kibo (centre) and Mawenzi (east) are aligned 30 km apart on an axis striking at 110°. Shira (2.5–1.9 Ma) is part of the Older Volcanism (Pliocene). Mawenzi (1.0–0.45 Ma) and Kibo (0.48–0.15 Ma) are part of the Younger Volcanism (Late Pleistocene–Holocene). A total volume of 4,790 km^3 has been estimated to have erupted: 500 km^3 from each of Shira and Mawenzi and 3,790 km^3 from Kibo. The dominant flow directions for the three centres are southwest (Shira), east (Mawenzi) and north, south and west (Kibo). Parts of Shira and Mawenzi are obscured by the younger Kibo volcanism and for this reason, as well as technical difficulties with the ascent of the latter, the older cones are poorly known. The parasitic cones and flows depicted on the map are probably part of the Kibo volcanism. Shira and Mawenzi produced mostly basalt and trachybasalt. Kibo displays a differentiation trend of increasingly silica undersaturated lavas: trachyandesite gives way initially to phonolite and then to nephelinite. Lahars or debris avalanche deposits (DADs; see Box 13.1) are associated with Mawenzi (an area of 1,000 km^2 within Kenya—not shown on the map) and Kibo, the latter occurring on the southern slopes.

12.4.1   Shira

Shira is located on the broad, western shoulder of Kibo. The highest point is the western rim of an eroded crater. Plazkegel is a prominent cone-shaped hill rising some 240 m above the floor of this crater, a few kilometres east of the ridge. This is an agglomerate plug which is associated with a radial dyke swarm. The crater is partly filled with younger sediments and the northern and eastern parts are covered by lava flows associated with Kibo.

12.4.2   Mawenzi

The precipitous cone of Mawenzi is capped by a horseshoe-shaped ridge, the remnants of an eroded crater. The summit ridge on the western and southern sides has near-vertical outer cliffs. Steep ravines on the outer slopes include two large features, the Greater and Lesser Barrancos, which are associated with collapse of the crater wall on the eastern side. This event triggered the Mawenzi DAD. Two main vents have been identified: Mawenzi and the Neumann Tower (located 2 km to the east of the summit). A number of intrusive gabbroic plugs occur. The ribbed appearance of the Mawenzi cone is ascribed to the erosion of a swarm of between 600 and 800 radial and concentric dykes.

12.4.3   Kibo

The upper part of Kibo at an elevation of between 4,500 and 5,638 m reveals an eroded caldera. This section of the mountain resembles an upturned basin (Fig. 12.3) perched on the broad outer shoulders of the volcano. The outer slopes of the caldera are particularly steep and constitute the most strenuous part of the ascent. The caldera event has an age of approximately 0.17 Ma. A scarp associated with the caldera fault has been located on the southern and western sides of the summit plateau, but may have been obscured by the icefields on the northern and eastern sides during the geological mapping (Fig. 12.4). The caldera is capped by an elliptical summit plateau that measures 2.7 km by 1.9 km. The caldera is tilted so that the highest point, Uhuru Peak, is located on the southern rim. In this part of the caldera, the inner walls rise 180 m above the summit plateau.

The caldera collapse triggered the Kibo DAD (or lahar) and carved the Barranco, a deep gorge with high cliffs located on the outer western slopes, prior to spreading onto the southern slopes. Most of the older lava flows on Kibo are only exposed on the lower, southwestern slopes as they are covered by younger flows that originated near the summit in other areas. Ten groups of lavas have been identified, most of which are separated by cycles of erosion. The older lavas are grouped together for simplicity on the map (Fig. 10.2). They include trachyandesite with a distinctive porphyritic texture (0.48 Ma), the Lent and Small-rhomb Porphyry Groups (0.36–0.38 Ma) and the Caldera Rim Group (0.23–0.17 Ma). The three latter groups are dominated by trachyte and phonolite. The scarp associated with the caldera includes beds of agglomerate and cinder. A body of syenite related to the Small-rhomb Porphyry Group and some 150 m in thickness, has intruded the outer rim of the caldera in the Barranco. This feature may be compared with the intrusive bodies of syenite that blocked the conduits of the Mount Kenya Volcano. The youngest lava flows on Kibo are intercalated with glacial deposits (see below).

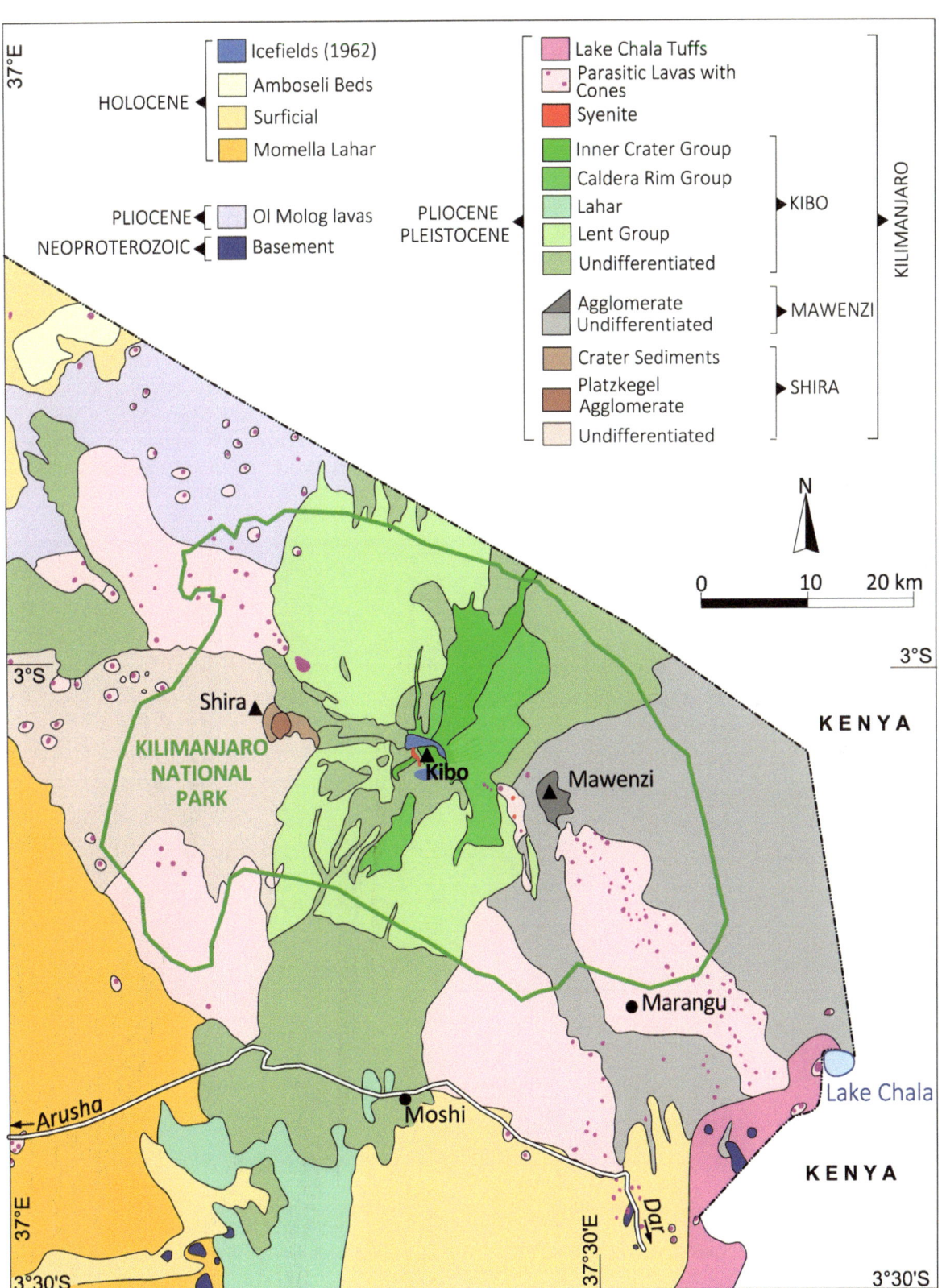

Fig. 12.2 Geological map of the Kilimanjaro massif simplified from Downie and Wilkinson (1972). The Small-rhomb Porphyry Group is included with the Lent Group

12.5   Reusch Crater

The 820-m-wide Reusch Crater (Inner Crater) is located asymmetrically in the northeastern part of the summit plateau (Fig. 12.4). The Inner Crater Group (0.15 Ma) which was erupted from the Reusch Crater includes a nepheline-bearing phonolite (within the crater) and a large flow of nephelinite that broached the caldera rim extending down the outer northeastern slopes (Fig. 12.2). The crater includes a small parasitic cone (Inner Cone), the walls of which are so severely eroded the altitude is 60 m lower than Uhuru Peak on the caldera rim. The Inner Cone in turn hosts two of the youngest features of the Kibo volcanism, the Ash Cone and Ash Pit. The Ash Pit is a central vent with a diameter of 340 m and depth of 130 m. This vent may originally have hosted a lava lake that withdrew so rapidly as to preserve the steep inner slopes (they become vertical with depth). Sulphur-rich fumaroles with temperatures of 77–104 °C and steam vents have been reported from the fractured western wall and from the 'Terrace'. Recent reports suggest these features are no longer active. Guest and Sampson (1952) remarked on the presence of a deposit of

approximately 6,500 tonnes of sulphur in the Reusch Crater, where it forms a thin (0.15 m) crust.

12.6   Parasitic Volcanism

Some 250 parasitic cones have been identified on the lower and intermediate slopes of Kilimanjaro (Fig. 12.2). They are dated at 0.20–0.15 Ma and may be contemporaneous with the caldera event and the Reusch Crater. They form 60–100 m high cones of cinder and ashes ascribed to Strombolian-type activity. The parasitic cones reveal a broad compositional range, including picrobasalt, trachybasalt, ankaramite and basanite. Lake Chala on the Kenyan border occurs in a crater from which an extensive calcareous tuff was ejected. These eruptions are thought to have destroyed a settlement several hundreds of years ago, providing the most recent evidence of volcanic activity although there are no definitive historical reports. The activity at Kilimanjaro may exhibit some reciprocity with Mount Meru which is located 50 km to the southwest (Wilkinson et al. 1986). The Momella Lahar, one of the world's largest DAD's derived

Fig. 12.3 Image of the upper slopes of Kibo showing the eroded caldera which includes the Furtwängler Glacier (FG), the last remaining ice sheet on the summit plateau. The Kersten and Decken glaciers are relics of slope glaciers in the Southern Ice Field. *Source* Google-Earth Image 2017, DigitalGlobe, CNES/Airbus

from the partial collapse of Mount Meru has spread onto the lower slopes of Kilimanjaro (Figs. 12.2 and 13.2).

12.7   Icefields

The distribution of ice on the three peaks of Kilimanjaro has changed repeatedly during and since the Late Pleistocene Ice ages (Cullen et al. 2013; Hardy 2011; Pepin et al. 2014;

Thompson et al. 2002). The half dozen or so early glaciations (680,000–130,000 BP) resulted in extensive ice sheets, as did the Main Ice Age (110,000–12,000 BP). During the Last Glacial Maximum (20,000 BP), the entire massif was covered by an icecap with an estimated area of 400 km^2 (Osmaston 2004). Mapping of glacial deposits and moraines intercalated with lava flows enabled Downie and Wilkinson (1972) to identify three of the older glaciations and the Main Ice Age on Kibo. (They are not shown on the map for reasons

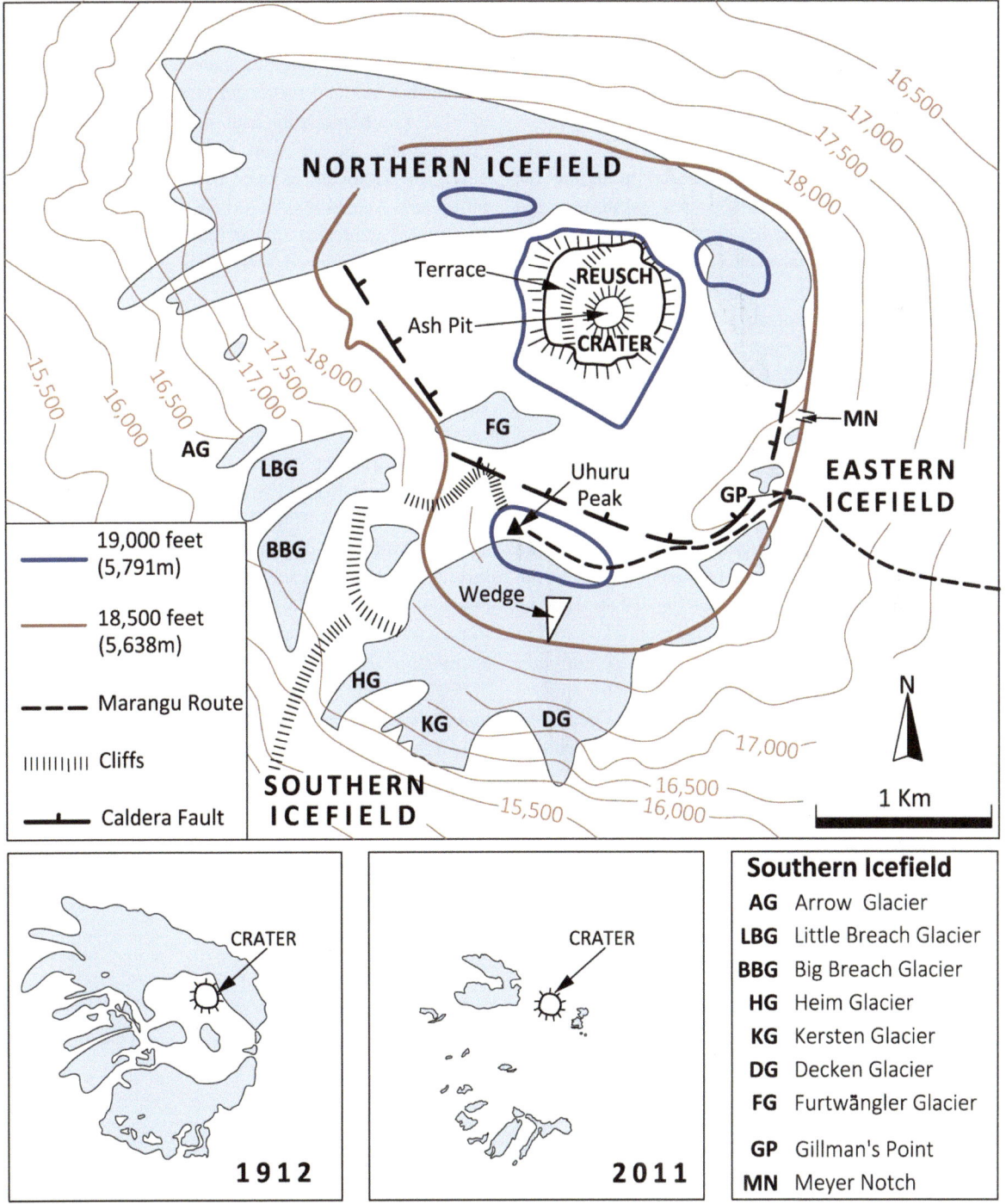

Fig. 12.4 Map of the Kibo summit plateau based on Mitchell (1971) with distribution of icefields in 1912 and 2011 from Hardy (2011). The outer rim of the caldera correlates approximately with the 18,500-foot contour

of scale). Extensive moraines were also mapped in the radial valleys on Mawenzi. The moraines on Kibo are typically some 5 m thick and 6 km in length. They are particularly well developed in valleys on the southern slopes at elevations of 3,350–3,960 m, where they were deposited by the enlarged southern glaciers. Small moraines located on Kibo are associated with ice movements during the Little Ice Age and Mini-Ice Age. Cores drilled into the Northern Ice Field on Kibo are an important record of climate change in East Africa (Sect. 6.3). They reveal the ice has a maximum age of 11,700 BP and is not a relict of a Late Pleistocene icecap. The Pleistocene icecap is presumed to have disappeared entirely around 12,000 BP due to an extremely dry period.

12.8 Holocene Recession

Slope glaciers were probably present on Kibo throughout the Holocene, but the summit icefield appeared and disappeared repeatedly, probably due to the dry climatic cycles described in Sect. 6.3. Early reports intimate Kibo was covered by a large icecap. The first ascent via the Hans Meyer Notch was made when the summit plateau was almost entirely ice covered. Only the Reusch Crater was ice free. This historical, late nineteenth-century icecap probably covered an area of 20 km^2, with slope glaciers extending to altitudes of 4,500 m. An aerial photograph revealed the historical icecap to have already shrunk considerably by 1938, to 11.4 km^2 (Plate 12.3a). A map of the summit plateau after Mitchell (1971) reveals three icefields with glaciers named after the early German explorers and geographers (Fig. 12.4a). The extent of the icecap at this time can be compared with the earliest (1912) and latest (2011) maps (Fig. 12.4b, c). The ice sheets and glaciers were considerably more extensive in the 1970s and 1980s (Plate 12.3b) than they are currently (Plate 12.4a). Only eight ice sheets and glaciers now survive with a combined area of 1.76 km^2. The Heim, Great Barranco and Little Penck are static ice sheets on the caldera rim. The Kersten, Decken, Rebmann and Credner are slope glaciers with minimum altitudes of approximately 4,900 m. The Northern Ice Field has the largest volume of the remaining glaciers. The Furtwängler is the last remaining ice sheet on the summit plateau, an Alpine desert devoid of vegetation (Plate 12.4b).

Recession of the ice on Kibo during the past 100 years has occurred at approximately 1% annually, with some 40% having disappeared since 1998. The possibility of an increased rate since the 1950s is discounted. Despite air temperatures at 5,000 m of −7 °C, the ice is melting due to solar radiation on exposed vertical walls. Most ice, it is predicted, will disappear by 2040, in part as surface areas are persistently shrinking. The absence of permanent ice from the Reusch Crater (even during the enlarged historical icefields) is evidence of geothermal heat (fumaroles and steam vents). Large parts of the summit plateau have a relatively high geothermal flux and meltwater with an average temperature of −1.2 °C is transporting heat into the glaciers. The potential disappearance of ice from Kibo would be unprecedented during the Holocene as even in the extreme drought of the First Dark Age (4,000–3,700 BP), some ice survived. The loss of ice may also in part be due to decreased rainfall resulting in part from deforestation of the lower slopes.

12.9 Trekking

Kilimanjaro provides an unusual opportunity to ascend a substantial peak with little technical difficulty. There are seven main trekking routes of which the most popular is the 5-day Marangu Route, which accesses the summit plateau via Gillman's Point (Fig. 12.3). The relative ease of ascent coupled with a rapid gain in altitude can, however, lead to serious problems and acclimatisation by ascending slowly is essential. The area below the Southern Ice Field and near the Breach and Arrow glaciers was reported as being subjected to severe avalanches in the 1950s. A fatal incident occurred in 2006 at the base of the Arrow Glacier. The entire mountain is volcanic and the loose scree slopes on the Kibo Caldera consist of severely eroded lavas and ashes, not shale as is often reported. Erosion is enhanced by the extreme diurnal ranges of temperature in which water freezes during the night.

Plate 12.2   a The gentle shoulders and immense size of the Kilimanjaro massif viewed from the northern side, Amboseli, Kenya; **b** View of the Saddle on the Marangu Route looking towards the outer walls of the Kibo Caldera

Plate 12.3   **a** Aerial photograph of the Kibo ice cap looking south shows the extensiveness of the historical ice cap, the raised southern rim of the caldera, and the Reusch Crater (1938); **b** Part of the Southern Ice Field, Kibo (1980)

Plate 12.4 **a** The Northern Ice Field on the summit plateau of Kibo. Photograph courtesy of Douglas Hardy, University of Massachusetts Geosciences (2012). **b** View looking north near Uhuru Peak, Kilimanjaro, with the Northern Ice Field (background) and relict of the Furtwangler Glacier located on the ash-strewn surface of the Reusch Crater (foreground). Photograph courtesy of Douglas Hardy, University of Massachusetts Geosciences (September 2017) who observed that the size of the ice area in the Furtwangler Glacier was some 32% less than two years previously

References

Cullen, N. J., Sirguey, P., Mölg, T., Kaser, G., Winkler, M., & Fitzsimons, S. J. (2013). A century of ice retreat on Kilimanjaro: the mapping reloaded. *The Cryosphere, 7*, 419–431.

Dawson, J. B. (2008). The Gregory Rift Valley and Neogene-Recent Volcanoes of Northern Tanzania. *Geological Society Memoir, 33*, 102 p.

Downie, C., & Wilkinson, W. H. (1972). *Geology of Kilimanjaro. Joint University of Sheffield-Tanzania Geological Survey Expedition* (253 p).

Guest, N. J., & Sampson, D. N. (1952). *Sulphur on Kibo, Kilimanjaro* (Mining Research Pamphlet 65). Geological Survey of Tanganyika.

Hardy, D. R. (2011). Kilimanjaro. In V. P. Singh, P. Singh, & U. M. Haritashya (Eds.), *Encyclopedia of Earth Sciences Series. Encyclopedia of snow, ice and glaciers* (pp. 672–679), Springer.

Jennings, D. J. (1971). Geology of Mount Kenya. In J. Mitchell (Ed.), *Guide book to Mount Kenya and Kilimanjaro* (pp. 30–41). Mountain Club of Kenya.

Mitchell, J. (1971). *Guide book to Mount Kenya and Kilimanjaro* (240 p). Mountain Club of Kenya.

Nonnotte, P., Guillou, H., Le Gall, B., Benoit, M., Cotten, J., & Scaillet, S. (2008). New K-Ar age determinations of Kilimanjaro volcano in the North Tanzanian diverging rift, East Africa. *Journal of Volcanology and Geothermal Research, 173*, 99–112.

Pepin, N. C., Duane, W. J., Schaefert, M., Pike, G., & Hardy, D. (2014). Measuring and remodelling the retreat of the summit icefields on Kilimanjaro, East Africa. *Arctic, Antarctic, and Alpine Research, 46*, 905–917.

Osmaston, H. (2004). Quaternary glaciations in the East Africa Mountains. In J. Ehlers & P. I. Gibbard (Eds.), *Developments in quaternary sciences, quaternary glaciations extent and chronology part III: South America, Asia, Africa. Australasia, Antarctica, 2C* (pp. 139–150). Amsterdam: Elsevier.

Thompson, L. G., Mosley-Thompson, E., Davis, M. E., Henderson, K. A., Brecher, H. H., Zagorodnov, V. S., et al. (2002). Kilimanjaro ice core records: Evidence of Holocene climate change in tropical Africa. *Science, 298*, 589–593.

Wilcockson, W. H. (1956). Preliminary notes on the geology of Kilimanjaro. *Geological Magazine, 93*, 218–228.

Wilkinson, P., Downie, C., Cattermole, P. J. & Mitchell, J. G. (1983). *Notes accompanying the quarter degree sheet 55: Arusha.* Geological Survey of Tanzania.

Wilkinson, P., Mitchell, J. G., Cattermole, P. J., & Downie, C. (1986). Volcanic Chronology of the Meru-Kilimanjaro region, Northern Tanzania. *Journal of the Geological Society of London, 143*, 601–605.

Williams, J. G., Arlott, N., & Fennessy, R. (1967). *Collins field guide to National Parks of East Africa* (336 p). Hong Kong: Harper Collins.

Arusha National Park (Mount Meru)

Abstract

The Arusha National Park, northern Tanzania is dominated by Mount Meru, which at 4,565 m is Africa's fourth highest summit. Meru is a giant stratovolcano, part of the Younger Volcanism located on older, faulted volcanic terranes in the northern Tanzania divergence. The main cone has a diameter of 25 km. It was built up from numerous, explosive, Plinian-style eruptions that occurred between 0.20 Ma and 80,000 BP. The most spectacular feature of the volcano is a horseshoe-shaped caldera with an estimated age of 7,800–7,000 BP. The western side of the caldera reveals sheer inner walls and is capped by a rocky summit ridge. The caldera contains the 1,067-m-high Ash Cone, an unconsolidated pyramid of ash and cinders, which is one of the youngest volcanic features of Meru. The collapse of the eastern sector of the cone produced a large debris avalanche deposit (DAD), the Momella event, which extends over 35 km onto the lower slopes of Kilimanjaro. The caldera collapse and Momella event can be compared with the catastrophic eruption of Mount St. Helens in 1980. The avalanche at Meru was far larger and carved out a distinctive undulating terrane that contains the Momella Lakes, important habitats for migrating birds. The montane forests that girdle the lower and central slopes of the mountain are particularly extensive and are refuges for large mammals and numerous species of birds. The slightly older Ngurdoto Volcano, which is situated in the southeastern arm of the park, includes a well-preserved summit crater protected from visitors. Meru is gazetted as an active volcano (the last activity was in 1910) and should be monitored as potentially hazardous, particularly in light of the explosive style of volcanism and proximity to the regional town of Arusha.

Keywords

Ash cone • Caldera • DADs • Meru • Momella Lakes • Sector collapse • Stratovolcano

Photographs not otherwise referenced are by the author.

Plate 13.1 The giant, horseshoe-shaped caldera at Mount Meru contains the Ash Cone which last erupted in 1910. The unvegetated flows in the foreground are also evidence of relatively recent activity although they have not been radiometrically dated

13.1 Introduction

Despite its relatively small size (137 km^2), the Arusha National Park is one of the most spectacular in East Africa. The dominant feature is Mount Meru, a giant stratovolcano (Cattermole 1982) that includes a horseshoe-shaped caldera with a largely unconsolidated cone of ash and cinder (Plate 13.1). The volcano has a diameter of 25 km and rises more than 3,500 m above the regional plateau to the north of the regional town of Arusha (Fig. 13.1). The montane forests that girdle the central and lower slopes are extensive and include stands of giant Podocarpus trees. The forests are an important habitat for large mammals and host numerous bird species. Most landforms in the park have been created and shaped by geological processes associated with explosive phases of volcanism and erosion during the Late Pleistocene–Holocene. Meru is capped by a summit ridge that leads to the rocky protuberance of Socialist Peak, which at 4,565 m is the fourth highest in Africa (Plate 13.2a). Many of the most significant features, including the caldera, ash cone and summit ridge are visible on a three-dimensional

satellite image (Fig. 13.2). Meru is gazetted as an active volcano as there was minor activity in the past century (Guest and Leedal 1956; Guest and Pickering 1966).

13.2 Mythology

The name Meru is derived from ancient Hindu and Buddhist religious scripts, in which a mythological mountain is described as a sacred place in the centre of our universe, in both a physical and spiritual sense. The concept of a mountain with seven rings, separated by water, is central to the mythology of several ancient cultures. The active Mediterranean volcano of Santorini with its sea-filled caldera enclosed by a ring of islands may in part fit these descriptions. Both Meru and Santorini experienced highly explosive, Plinian-style eruptions that included caldera events. Multiple caldera events can occur and seven such events have been recognised at Santorini (Druitt et al. 1999). Volcanism is typically rejuvenated in the centre of calderas and can form new cones, such as the active Nea Kameni Volcano, a newly formed island in Santorini, and the Ash Cone at Meru.

13.3 Older Volcanic Terranes

The Meru Volcano is built upon an older, faulted, volcanic terrane within the structurally-complex northern Tanzania divergence (Fig. 5.4). The region includes two principal fabrics, the northwest–southeast trending Lembolos Graben and the north–south trending Oljoro Graben. The volcanic rocks in the vicinity of Meru may have arisen from magmas fed through weakness associated with these grabens (Dawson 2008). Some of the Pliocene and Early Pleistocene volcanism that pre-dates Meru, i.e. the Older Volcanism, is shown on the simplified geological map (Fig. 13.3). The latter includes the phonolite lavas of the Meru West Group (3.1–2.4 Ma), the alkali basalt-phonolite lavas of the Oljoro Graben (2.5–2.3 Ma), and the flows of alkali basalt-phonolite on the lower, western slopes of Meru (1.5 Ma). Several volcanic centres which are part of Younger Volcanism (Late Pleistocene) and yet pre-date Meru are also identified. Little Meru (0.40–0.30 Ma) is a subsidiary peak on the northeastern slopes of the main cone comprised of explosive nephelinite breccias with clasts of older material (including phonolite) which may have been derived from Meru West (Wilkinson et al. 1986). The Ngurdoto Volcano is protected by an arm of the Arusha National Park that projects southeast from Meru. This severely eroded cone includes a well-preserved summit crater with a diameter of 4 km by 3 km and a depth of 360 m. Ngurdoto is associated with alkaline silicate (nephelinite) and natrocarbonatite lavas (Roberts 2002) and although it has not been dated is older than Meru, as is the adjacent Matuffa Crater.

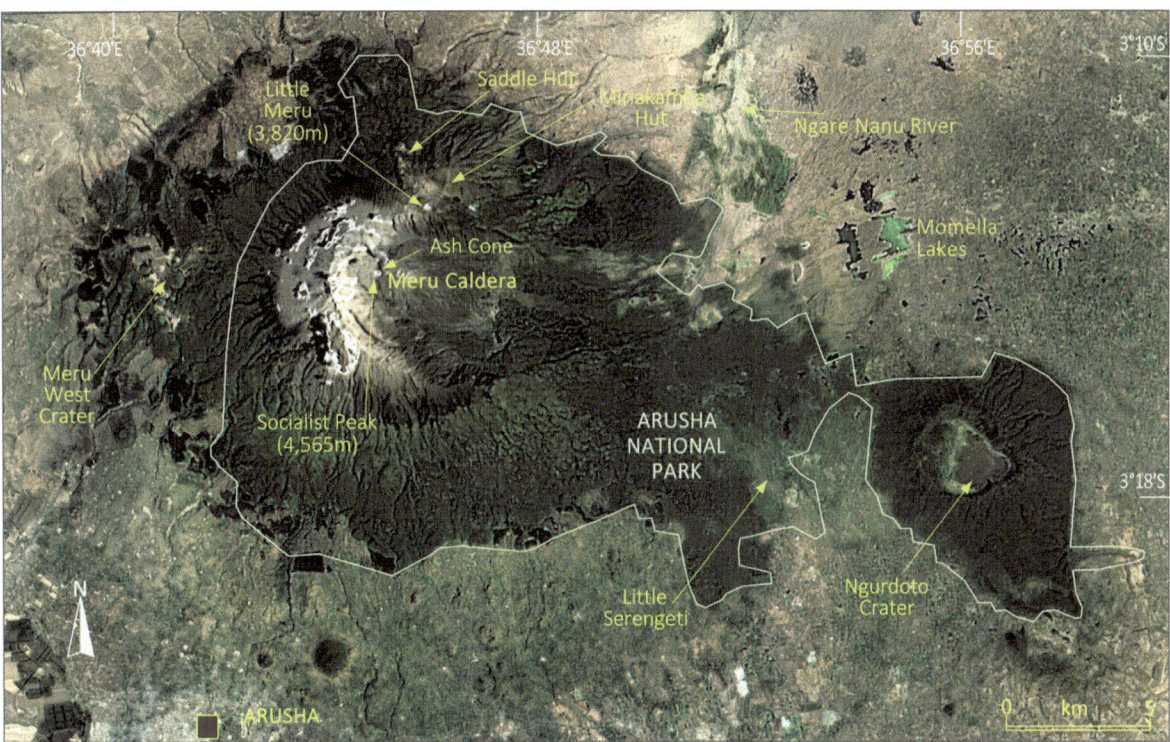

Fig. 13.1 The Arusha National Park is dominated by the giant cone of Mount Meru. *Source* Google-Earth Image 2017, DigitalGlobe, CNES/Airbus

Fig. 13.2 Three-dimensional satellite image looking northwest over Mount Meru towards the older Monduli and Tarosera volcanoes. Meru includes an annular ring of forest on the lower slopes and a massive, partially collapsed caldera, open to the east, with a central cone of ash and cinder. Little Meru is visible as a small triangular-shaped cone on the northeastern slopes. The Ngurdoto Crater occurs in the foreground. The hummocky ground and lakes associated with the Momella DAD occur east of the Meru Caldera and north of the Ngurdoto Crater. *Source* NASA Landsat 7 ETM+ image mosaic for the year 2000 from the University of Maryland Global Land Cover Facility, overlaid on elevation data from the Shuttle Radar Topography Mission, courtesy of Philip Eales, Planetary Visions

13.4 Main Volcanism

Volcanism associated with the main phase of activity at Meru peaked at 0.20 Ma–80,000 BP, after Little Meru had become extinct (Wilkinson et al. 1986). This coincided with the waning of activity in the Kibo Volcano (Kilimanjaro), an indication these two giant volcanoes may share a common plumbing system. The Main Cone Group is the most extensive of the volcanic subdivisions recognised at Meru (Fig. 13.3). This group is dominated by phonolite breccia and tephra that formed from repeated Plinian-style eruptions, a style of volcanism that characterises continental rifts (Sect. 5.10). Formation of the main cone (which may originally have attained a height of over 5,000 m) was followed by a period of quiescence with cycles of intense erosion occurring during the Main Ice Age. Numerous deep gullies and ridges developed during this period.

At approximately 60,000 BP, a new burst of activity at Meru resulted in formation of the Summit Group (Fig. 13.3). This unit of phonolite and nephelinite lava flows forms the summit ridge with its prominent 150-m-high rock dome (Plate 13.2b). The phonolite lavas may contain phenocrysts of alkali feldspar. Deposits of fine- and coarse-grained ash are intercalated with the lava flows on the summit ridge (Plate 13.3a). Formation of the Summit Group was followed by a renewed period of quiescence, albeit with extensive parasitic activity. Parasitic features are particularly well developed on the lower, southern slopes. Some parasitic craters are partially infilled by lakes, many of which are very scenic, e.g. the Duluti Crater at an elevation of 1,300 m, located 13 km to the east of Arusha.

The parasitic activity was disrupted by Plinian-style events with the eruption of pyroclastic flows high on the flanks of the main cone and with the formation of extensive deposits of pumice and ash on the outer slopes. These deposits mapped as the Mantling Ash blanket large areas of the northern, western and southern slopes (Fig. 13.3). They are well exposed in some river beds, particularly on the western slopes, where they may be as much as 20 m in thickness (Vye-Brown et al. 2014). The ash columns associated with these relatively young Plinian eruptions could have attained heights of 23 km, similar to those described by Pliny the younger in conjunction with the 79AD eruption of Vesuvius (Box 5.1).

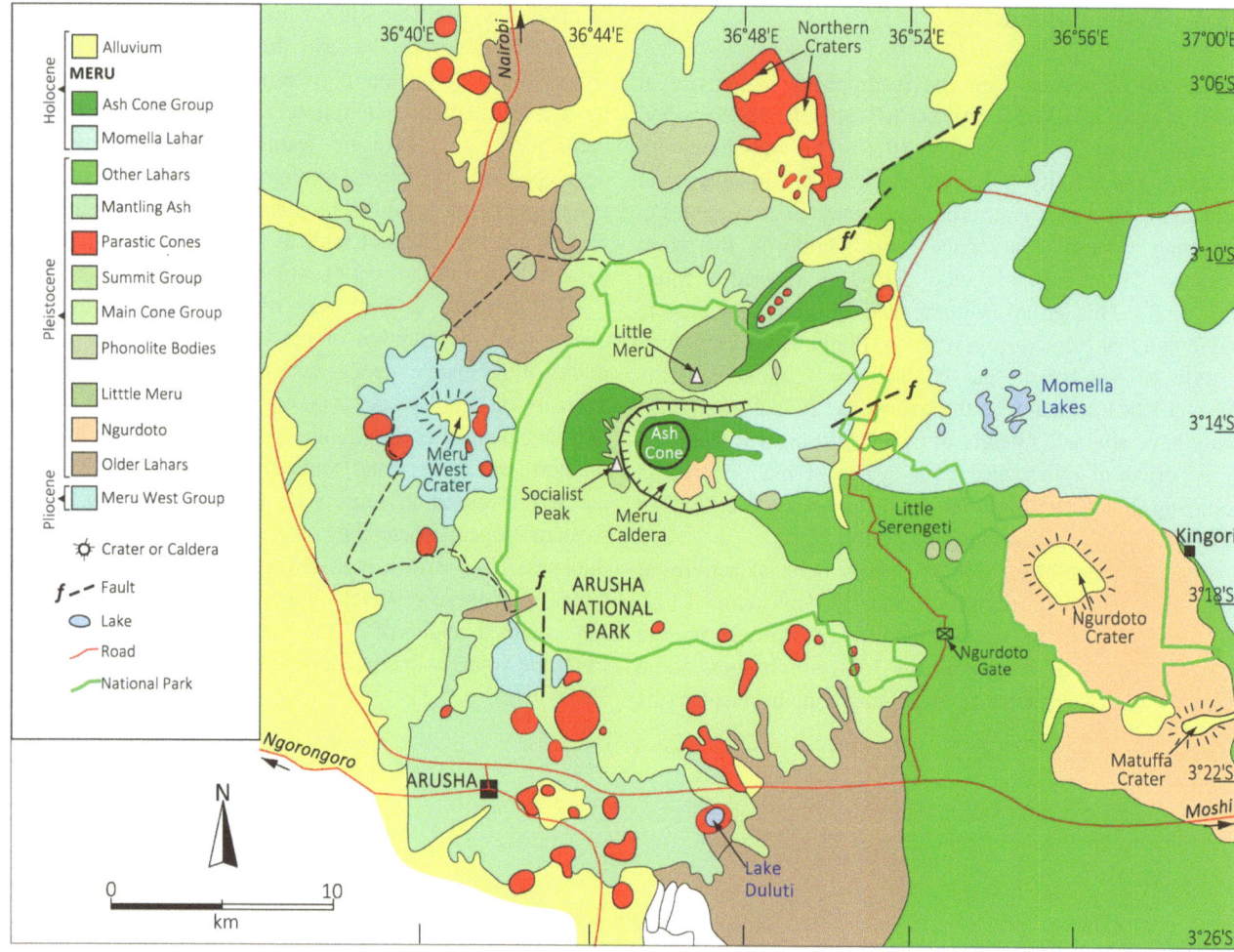

Fig. 13.3 Geological map of the Arusha National Park and surrounding area simplified from the Geological Survey of Tanzania 1:125,000 quarter degree sheet 55. The original terminology, relative ages and mapping units of Wilkinson et al. (1983) are retained, e.g. use of lahar rather than DAD. Recent studies suggest the Summit Group may be younger than the Momella Lahar (or DAD)

13.5 Sector Collapse and the Momella DAD

After the persistent activity of the Late Pleistocene, the Holocene saw a resurgence of catastrophic volcanism at Meru. The eastern sector of the main cone disintegrated at approximately 7,800–7,000 BP to create the horseshoe-shaped caldera (Wilkinson et al. 1986). The caldera has a length of 8 km, width of 5 km and is entirely open to the east (Fig. 13.2). The 1,300-m-high, near-vertical internal walls on the western face include numerous layers of lavas and ash, as well as evidence of block faulting, and feeder dykes (Plate 13.3b). Formation of the caldera and partial collapse of the cone at Meru can be compared with the 1980 eruption of Mount St. Helens (Box 13.1), as originally suggested by Roberts (2002). This would have involved a far greater explosive force than at Mount St. Helens as the caldera and cone are far larger. Major seismic shocks would have preceded this event. The collapse of entire sectors of volcanic cones results in debris avalanche deposits (DADs), as discussed in connection with Meru by Delcamp et al. (2015). The lahars of the earlier mapping (Fig. 13.3) can be reinterpreted as DADs.

The Momella DAD is associated with the sector collapse that created the caldera. This feature was originally mapped as extending over 35 km to the lower slopes of Mount Kilimanjaro, and with a surface area of approximately 400 km^2 (Fig. 12.3). Recent studies have suggested this deposit is far larger and can be conjoined with lahars that were originally mapped as separate, older features. The areal extent of the Momella DAD is now estimated as approximately 1,249 km^2 and the volume is calculated as 18 km^3, constituting one of the largest DADs ever recorded (Delcamp et al. 2015). DADs create characteristic undulating terranes, as the fast-travelling avalanches gouge out the land surface (Plate 13.4a). The Momella DAD includes house-sized boulders of phonolite derived from the main cone.

The Momella DAD formed during the hot and humid climatic regime of the Early Holocene, and some parts of the deposit have been redistributed by fluvial activity into fan and fluvio-volcanic sequences. The lateral blast or surge associated with the sector collapse would have affected a far larger area than the caldera or avalanche and most of the forest over many tens of km^2 would have been destroyed. The blast would have had catastrophic effects on the early inhabitants of the region, such as the Hadzabe tribe. The sector collapse that produced the Momella DAD may also have created a thin layer of ash that mantles large parts of the surviving cone (Delcamp et al. 2015). These events may be correlated with formation of a small lava dome and nephelinite flows within the caldera. One of the nephelinite flows contains xenoliths that may have been transported from the mantle (Roberts 2002).

Box 13.1: Meru and Mount St. Helens

*Formation of the caldera and DADs at Meru may be compared with the 1980 eruption of Mount St. Helens, in Washington State, USA. This Plinian-style eruption was triggered by the release of pent-up pressure within the cone associated with the expansion of the magma chamber. The eruption was preceded by seismic shocks including a magnitude 5.1 event. Most of the northern sector of the cone collapsed, or blow out, **prior** to the extrusion of air-fall ash and pyroclastic flows (Glicken 1996). The eruptive products included fast-travelling pyroclastic flows, comprised of hot ash, pumice and gas (they were restricted to a small fan-shaped area on the upper slopes), as well as a huge ash plume that rose to a height of almost 20 km. The abrupt collapse of part of the Mount St. Helens cone triggered a debris avalanche of rock, ash and hot gases to rush down the northern flanks. This avalanche, together with the near-instantaneously formed mudflows and floods was highly destructive and travelled some 27 km. A feature that was not appreciated prior to the Mount St. Helens*

event, however, is that pressure release by the rapid disintegration of the cone triggered an instantaneous expansion (explosion) of high temperature–high pressure steam within the magma chamber. This expansion created a lateral blast or basal surge which rapidly overtook the avalanche and, travelling at speeds of up to 1078 km/h devastated the surrounding landscapes. The lateral blast and debris avalanche at Mount St. Helens caused far more intensive and widespread damage than the eruptive flows and ashfall (Fig. 13.4).

13.6 Momella Lakes

The three large lakes and smaller lakes, ponds and marshes in the northeastern segment of the park, occur in areas of hummocky ground associated with the Momella DAD. The lakes fill hollows in the debris deposit and can be assumed to have a similar maximum age, i.e. approximately 7,800–7,000 BP. Big Momella Lake is the deepest of the lakes (10–30 m) and is moderately alkaline. The more scenic Small Momella Lake is shallower (4–10 m) and although alkaline and salty in the central parts, includes freshwater sections in which Hippopotamus and aquatic birds thrive. The smaller Rishateni Lake is unusually rich in dissolved fluorine, possibly the highest ever recorded in natural lakes. The fluorine is derived from the erosion of the alkaline-rich volcanic rocks. The water used for domestic purposes in the local villages, and also in Arusha may similarly have anomalously high contents of fluorine. Both the Big Momella and Rishateni Lakes contain sufficient cyanobacteria for migrating flamingo (Lihepanyama 2016) (Plate 13.4b). Flamingo may also occur on the small alkaline lakes of Elkekhotoito, Jembamba and Tululusia.

At approximately 1,800 BP, minor seismic activity caused the course of the Ngare Nanyuki River that drains the eastern slopes of Meru to change in a northerly direction. As a result, most of the Momella Lakes are now fed by groundwater within the porous debris deposits and by limited surface run-off and precipitation. Only the Small Momella Lake remains part of an underground river system; the Large Momella and Rishateni Lakes are mostly stagnant. The Momella Lakes are important stopovers for a large variety of birds that migrate between Europe/East Africa and southern Africa.

Fig. 13.4 View of Mount St. Helens and surrounding area (width of view approximately 150 km) looking south-east using elevation data produced by the Shuttle Radar Topography Mission in 2000. The explosive eruption of 18 May 1980, caused the upper 400 m of the mountain to collapse, slide and spread northward, covering much of the adjacent terrane (lower left). The distinctive, shortened form of the cone with its summit crater can be compared with the more typical triangular peaks of Mount Adam and Mount Hood (background left and right, respectively). The devastation caused by the avalanche and blast is still apparent 20 years later. The high rainfall has led to the substantial erosion of the poorly consolidated landslide material. The colour coding is related to topographic height (green at lower elevations, rising through yellow and tan, to white at the higher elevations). A similar scene of devastation can be presumed to have resulted from the sector collapse and Momella DAD at Meru more than 7,000 years ago. *Source* NASA

13.7   Ash Cone

The giant pyramid-shaped body of ash and cinder, known as the Ash Cone rises 1,067 m above the floor of the caldera in the northwestern corner (cover). This feature is correlated with the most recent activity at Meru, i.e. after the sector collapse. In 1910, small amounts of ash were ejected from the Ash Cone for a few days. Up until 1954, fumaroles were recorded in the ash cone, but in 1974 a survey revealed no activity. The unvegetated lava flows on the high, northeastern flanks of the Ash Cone may have erupted as recently as 1877, although there is no consensus or accurate dating of these events. The central parts of the caldera include small seasonal lakes (pans), recorded incorrectly as craters on tourist maps, that dry to reveal deposits of alluvium and salts.

13.8   Ecosystems

The Arusha National Park has been described as one of the 'hidden gems' of Africa, as it has much to offer and yet receives far less visitors than Kilimanjaro and the more famous parks farther west. The biodiversity conservation and equitable management of natural resources in this area are hampered due to the relatively large population within the rural community (Istituto Oikos 2011). Mount Meru has a important role in ensuring climate stability and water supply for a large area, including a rapidly growing urban population based on the regional centre of Arusha. A fundamental principle is the protection of the montane forests and fertile foothills of Meru, large parts of which occur outside of the park.

13.9 Tourism

The highlight of a visit to the Arusha National Park is a 4-day trek to the summit with overnights at the Miriakamba and Machame Huts (Fig. 13.1 and Plate 13.5a). Views of Kilimanjaro towards the east are an additional reward of this trek (Plate 12.1). The diversion to Little Meru is an important component of the ascent, as it helps offset altitude sickness and also provides fine views to the north, including that of the Oldoinyo Lengai Volcano. The final part of the ascent is undertaken almost entirely on the rocky summit ridge (Plate 13.2a), and this is both far less strenuous, and yet more exposed, than the crowded treks on Kilimanjaro. Sunrise observed from Socialist Peak reveals a triangular shadow derived from Kilimanjaro (Plate 13.5b). The slopes above approximately 4,000 m are devoid of vegetation, whereas the lower and central slopes and parts of the caldera floor are thickly forested. Hiking in these pristine montane forests with the abundant large game, including African Elephant and Cape Buffalo, a large variety of birds (>400 species), and scenic waterfalls, is an additional experience. The relative youthfulness of the forests, i.e. they would have had to regenerate after the blast associated with the sector collapse is intriguing. The 'Little Serengeti' on the lower, eastern slopes is an area of savannah grassland that supports a range of grazers. The Ngurdoto Volcano includes thickly forested slopes with a grassy crater floor which appears as an idyllic 'Garden of Eden' (Plate 13.5c).

(a)

(b)

Plate 13.2 a View of the summit ridge of Meru looking south includes gently dipping layers of well-bedded lavas and ashes with the rocky, dome-shaped outcrop of Socialist Peak visible in the background; **b** The resistant rock dome of Socialist Peak is comprised of lavas of the Summit Group, one of the youngest components of the volcano

Plate 13.3 **a** The flanks of the Meru Caldera include inter-layered deposits of fine- and coarse-grained ash; **b** The internal western wall of the Meru Caldera is built up of multiple beds of lava and ash. The prominent contact in the upper part of the face may be related to the Momella event

(a)

(b)

Plate 13.4 **a** The hummocky ground on the lower eastern slopes of Mount Meru is associated with the Momella DAD; **b** Greater and Lesser Flamingo fringe Big Momella Lake with hummocky ground typical of the debris avalanche deposit also visible

Plate 13.5 **a** The sheer western wall of the Meru Caldera glows in the early morning light behind the Miriakamba Huts; **b** Sunrise on the summit of Meru casts a triangular shadow of Kilimanjaro; **c** The Ngurdoto Crater provides a glimpse into a verdant 'lost world', as access is denied to this park-within-a-park

References

Cattermole, P. (1982). Meru—A Rift Valley giant. *Volcano News, 11,* 1–3.

Dawson, J. B. (2008). The Gregory Rift Valley and Neogene-recent volcanoes of northern Tanzania. *Geological Society London Memoir, 33,* 102 p.

Delcamp, A., Delvaux, D., Kwelwa, S., Macheyeki, A., & Kervyn, M. (2015). Sector collapse events at volcanoes in the North Tanzanian divergence zone and their implications for regional tectonics. *Geological Society of America Bulletin, 128,* 169–186.

Druitt, T. H., Edward, L., Mellors, R. M., Pyle, D. M., Sparks, R. S. J., Lanphere, M., Davies, M., & Barreiro, B. (1999). Santorini Volcano. *Geological Society London Memoir, 19,* 169 p.

Glicken, H. (1996). Rockslide-Debris avalanche of May 18, 1980, Mount St. Helens Volcano, Washington. *US Geological Survey Open-File Report 96–677,* 90 p.

Guest, N. J., & Leedal, G. P. (1956). *The volcanic activity of Mount Meru* (Records 3, 40-46). Geological Survey of Tanganyika.

Guest, N.J., & Pickering, R. (1966). *Notes accompanying quarter degree sheet 40: Gelai and Ketumbeine.* Mineral Resources Division of Tanzania.

Istituto Oikos. (2011). *The Mount Meru challenge: Integrating conservation and development in northern Tanzania.* Milano: Ancora Libri (Italy), 69p.

Lihepanyama, D. G. (2016). *Ecology of Lesser Flamingos in them Momella lakes, Arusha National Park, Tanzania.* Unpublished B.Sc. thesis, University of Dar-es-Salaam, 87 p.

Roberts, M. A. (2002). *The geochemical and volcanological evolution of the Mt Meru region, northern Tanzania.* Unpublished Ph.D. thesis, University of Cambridge.

Vye-Brown, C., Crummy, J., Smith, K., Mruma, A., & Kabelwa, H. (2014). Volcanic hazards in Tanzania. In: *British Geological Survey Open File Report OR/14/005*, 29 p.

Wilkinson, P., Downie, C., Cattermole, P. J. & Mitchell, J.G. (1983). *Notes accompanying quarter degree sheet 55: Meru.* Geological Survey of Tanzania.

Wilkinson, P., Mitchell, J. G., Cattermole, P. J., & Downie, C. (1986). Volcanic chronology of the Meru-Kilimanjaro region, Northern Tanzania. *Journal of the Geological Society of London, 143*, 601–605.

Amboseli, Chyulu Hills and Tsavo West National Parks

14

Abstract

The Amboseli and Tsavo West National Parks, cover vast areas of mostly semi-arid, scrub-covered plateaus in a part of southeastern Kenya that has remained largely unaffected by human impact. The Chyulu Hills is a newly proclaimed national park that abuts against the western boundary of Tsavo West. This contains one of the youngest volcanic ranges on Earth (Pleistocene–Holocene). Views of Kilimanjaro are an additional highlight of visits to these parks and Amboseli includes parasitic cones related to Kibo, the youngest component of this giant volcano. Despite being located some distance from the Gregory Rift, the volcanism of these areas is associated with the East African Rift System (EARS). The Chyulu Hills volcanism is characterised by individual and coalesced cinder cones, which typically rise some 900 m above the regional plateau. This volcanism has encroached into Tsavo West where barren Holocene-age features, including the Shaitani and Chaimu flows and cones with an estimated age of 1865–1866, can be distinguished from the vegetated Pleistocene lava fields. The Shaitani event includes a lava tube, which can be entered from areas of surface collapse. An intriguing feature of Tsavo West is the site of recent lavas unconformably overlying granite of the Neoproterozoic Mozambique Belt. The Amboseli and Tsavo West National Parks support large concentrations of wildlife despite the relatively arid terranes. Wildlife is sustained throughout prolonged dry seasons by underground water fed through porous rocks of the volcanic uplands. Swamps on the Amboseli Plains rely on groundwater from the Kilimanjaro massif. Tsavo West is fed by the Mzima Springs with water derived from the Chyulu Hills. The springs have created a small oasis, where huge volumes of underground water well up through the Pleistocene age Mzima lavas.

Keywords

Holocene volcanism • Lava tube • Mozambique belt Mzima springs • Regional plateau • Shaitani lava

---

Photographs not otherwise referenced are by the author.

© Springer International Publishing AG, part of Springer Nature 2018
R. N. Scoon, *Geology of National Parks of Central/Southern Kenya and Northern Tanzania*, https://doi.org/10.1007/978-3-319-73785-0_14

Plate 14.1 The source of the Shaitani lava (1865–1866) can be traced to a crater and cinder cone in the Chyulu Hills (back right). The basaltic flow reveals a central ridge and cinder-dominated flanks

14.1 Introduction

The Amboseli and Tsavo West National Parks in southeastern Kenya are visited by large numbers of tourists every year, many going to observe huge concentrations of wildlife that include some of the largest herds remaining of African Elephant. Both parks are dominated by extensive regional plateaus, semi-arid areas which extend to the coastal escarpment. The Chyulu (or Kyulu) Hills is a relatively newly proclaimed national park adjoining Tsavo West which protects a linear array of volcanic cones that form one of the youngest ranges on Earth. The Chyulu Hills volcanism is gazetted as an active system and encroaches into Tsavo West. Sparsely vegetated cinder cones and lava flows are one of the more significant geological features of this area (Plate 14.1). The contrast between the barren lava flows and open woodlands and bush-covered plains is marked. A satellite image reveals the principal features of this area, including the proximity of the huge massif of Kilimanjaro (in northern Tanzania), the dry lake bed and salt flats of Lake Amboseli, and the comparatively lush Chyulu Hills (Fig. 14.1). The view of Kilimanjaro to the south of the Amboseli National Park is one of the most spectacular in Africa.

The question as to how such large numbers of wildlife survive in these semi-arid environments was first raised by Thomson (1887). The answer is the unusual geology. The northeast monsoon, sweeping off the Indian Ocean produces localised, orographic rainfall on the volcanic uplands. The rainwater soaks through the porous volcanic rocks prior to emerging as underground springs on the regional plateaus. Without rifting and formation of volcanoes as part of the EARS, the regional plateaus would be too arid to support such large concentrations of wildlife.

14.2 Regional Geology

The geology of this part of Kenya was first described by J. W. Gregory, whilst travelling from Mombasa to the rift valley (Gregory 1896). When he camped at Kibwezi, Gregory examined the ropy-textured lavas that dominate the Chyulu Hills. He also visited the remarkable Yatta Plateau which he interpreted as a single lava flow exploiting a depression in the basement complex (Fig. 5.2). The geological map presented here is very schematic (Fig. 14.2), although it includes part of the Kibwezi area which was described and mapped in detail by Saggerson (1963). Many

Fig. 14.1 110-km-wide satellite image of part of southeastern Kenya and northern Tanzania including the Amboseli, Chyulu Hills and Tsavo West National Parks. *Source* NASA Landsat 7 ETM + image mosaic for the year 2000 from the University of Maryland Global Land Cover Facility, courtesy of Philip Eales, Planetary Visions

of the geological features in the area can be accessed from the roads connecting Mtito Andei and Tsavo to Oloitokitok and Amboseli.

Two principal groups of rocks are identified with the region, the Mozambique Belt (Neoproterozoic) and the volcanic and sedimentary rocks related to the EARS.

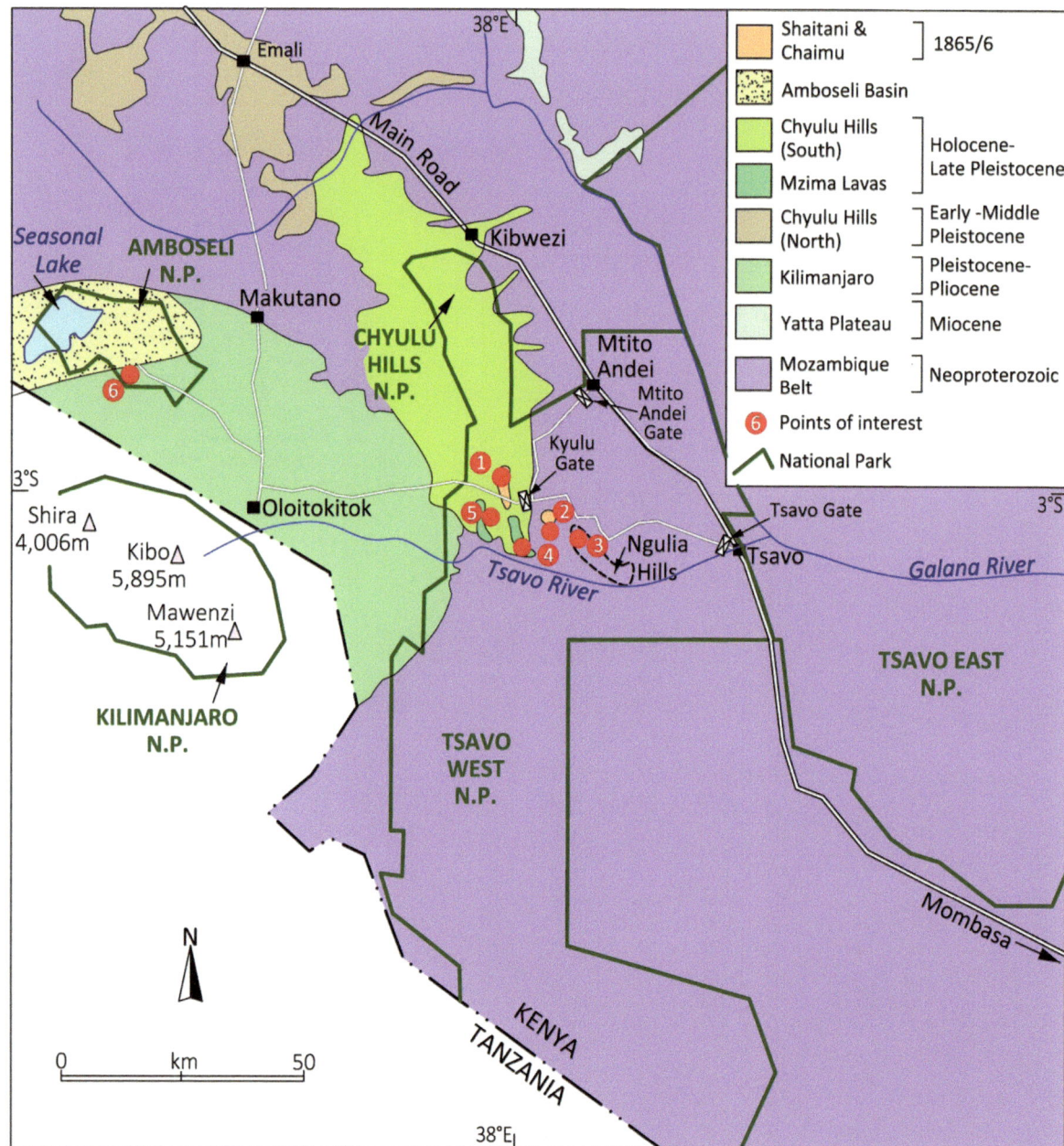

Fig. 14.2 Simplified geological map of the Amboseli, Chyulu Hills and Tsavo West National Parks. Areas of interest include: (1) Shaitani lava flow; (2) Chaimu cinder cone and lava flow; (3) Rhino Valley; (4) Mzima Springs; (5) Poacher's Lookout and (6) Noomotio Hill

Volcanism is remarkably prevalent on the Eastern Rift Platform and extends several hundred kilometres from the Gregory Rift Valley. There are many hundreds of separate volcanic cones and feeder vents within the Chyulu Hills and Saggerson (1963) recognised more than 350 in the Kibwezi area alone. Sedimentary sequences which include reworked volcanic ashes and volcaniclastic deposits, accumulated in warps on the regional plateau.

14.3   Amboseli National Park

Amboseli was first recognised as a wildlife reserve in 1906 and by 1974 an area of 392 km² was gazetted as a National Park. In 1991, the area was reclassified, together with adjacent lands where the nomadic Maasai have lived harmoniously for many years with the wildlife, as an

International Biosphere Reserve. The arid regional plateau appears as a rather desolate environment in the extensive dry seasons, but closer inspection reveals the presence of small swamps, areas of lush green, such as Narok and Enkong. The geology of Amboseli is rather poorly known due to the paucity of the outcrop. Most of the park is underlain by volcanic rocks associated with the Kibo component of Kilimanjaro (0.48–0.15 Ma). Some lava flows extend down the northern slopes onto the Amboseli Plains. The Kibo lavas and ashes have partially covered the earlier volcanism associated with Shira (2.5–1.9 Ma) and Mawenzi (1.0–0.45 Ma). The Kibo volcanism is best observed in localities where resistant parasitic cones, such as Noomotio Hill, emerge from the plateau. A short trail here with views of the lush swamps (Plate 14.2a) includes excellent information boards with details of the geology and ecology. The western section of the park includes part of Lake Amboseli, an ephemeral lake (which is invariably dry), with extensive salt flats. The lake was formed by the damming of palaeo-channels of the Pangani River caused by eruptions of Kibo. Lake Amboseli occurs in a shallow warp on the regional plateau that was partially infilled by volcanic ash and wind-blown sediments. The lake is rimmed by low, calcrete ridges (Plate 14.2b).

14.3.1 Environmental Changes

Some remarkable environmental changes have occurred at Amboseli during the Late Pleistocene and Holocene. Prior to eruption and formation of the Kibo Volcano, the Amboseli Plains were well wooded and swamps such as Narok and Enkong were forested. These areas probably supported far more diverse wildlife than currently seen. Two possible reasons for the changes have been suggested, a decrease in rainfall and a rise in the height of the water table, the latter having increased the salinity of soils. Most species of wildlife decreased during these changes with the exception of the African Elephant which not only prospered but transformed the ecosystem by enhancing deforestation. A similar, relatively minor event on a geological timescale occurred in Tsavo West during the extreme droughts of the 1980s and 1990s when elephant triggered extensive changes that included creating near-desert conditions by overgrazing.

The negative influence to tourism of huge numbers of elephant dying and the Dust Bowl conditions of the time were subsequently offset by the rejuvenation of grasslands and the positive spinoff of much larger numbers of grazers in an environment more pleasing to wildlife tourism. The possibility that the elephant can convert scrubland or woodland savannah into grasslands is now quite widely accepted. An important part of the Amboseli biosphere is

that the elephant herds, which are too large to be sustained wholly within the park, are now able to migrate within the perimeter areas with the support of the Maasai residents.

14.3.2 Groundwater

The swamps at Amboseli are fed by groundwater derived from the Kilimanjaro massif. Rainwater falling on the porous volcanic rocks is trapped at relatively shallow depths by layers of impermeable lavas and intrusive sills. The outward flow of groundwater is driven by the enormous hydrostatic pressure. The main source of water accrues during the rainy seasons (April–May and October–November), with additional supplies from snowmelt during the intervening months. The retreat of the icefields on Kibo (Sect. 12.5) is a major concern for the potential water supply. Amboseli is a delicate environment and the volcanic soils are powdery and contain few nutrients. They are regularly whipped up into spectacular dust storms with small tornadoes or dust devils.

14.4 Chyulu Hills National Park

The 150-km-long Chyulu Hills is comprised almost entirely of Pleistocene–Holocene volcanic rocks. In its entirety, the Chyulu Hills Volcanic Province covers an area of some 2,840 km^2, one of the most significant volcanic fields in East Africa. This province is generally divided into two sections, an older northwestern area where activity commenced in the Pleistocene (oldest age of 1.4 Ma) and a younger southeastern area that is dominantly Holocene (Spath et al. 2000). The northwestern area contains discrete eroded cones and vents, but the southeast reveals a more-or-less contiguous range of prominent cones. Individual and coalesced cinder cones in the southeastern area, as viewed from Tsavo West typically rise some 900 m above the regional plateau (Plate 14.3a). Many cones are aligned along either northwest trending Pleistocene age faults, or north–northeast trending Holocene-age fissures (Haug and Strecker 1995). Cones in the southeastern area include lava flows, which can be traced back to discrete vents (Plate 14.1). The older Pleistocene flows in the Chyulu Hills are dominated by foidite, with the younger Holocene activity consisting largely of basanite and alkali basalt. This increase in the silica content of the magmas over time is an intriguing feature, as this trend is the reverse of fractionation paths found in other magmas in the EARS. It has been suggested, the Chyulu Hills volcanism is an analogue for the very early part of continental rifting when melts are generated by the interaction between a mantle plume and a refractory lithosphere (Spath et al. 2001).

14.5 Tsavo West National Park

The Tsavo West and much larger and more remote Tsavo East National Parks, when combined, constitute the largest park in East Africa (over 21,000 km^2). This contribution is restricted to Tsavo West, which is separated from Tsavo East by the Mombasa-Nairobi road and railway line (Fig. 14.2). The area is known internationally for historical events including scenes in the film 'Out of Africa' (based on the book by Karen Blixen) and the 'Man Eaters of Tsavo', the latter being an account of the damage caused by two lions during the building of the railway in 1898, as documented by Colonel J. H. Patterson. In recognition of this latter event, as well as the presence of volcanic lava flows, Tsavo West is known as an 'Ancient Land of Lions and Lava' (Plate 14.3b). Tsavo West consists largely of undulating, scrub-covered hills and open plains. The Ngulia Hills are areas of grassland and scrubby bush, with rather poorly developed montane forest on the upper parts. The park is drained by the Galana and Tsavo Rivers.

14.5.1 Mozambique Belt

Large parts of Tsavo West are underlain by the Mozambique Belt, part of the crystalline Basement rocks associated with the major Neoproterozoic collision event that affected all of East Africa (Shackleton 1986). The principal rock types are granite, mica schist and marble (Mosley 1993). The more resistant lithologies form prominent hills, including insel-bergs, which penetrate above the recent covering of rift-related volcanic lavas and ashes. The Ngulia Hills is dominated by a large body of distinctive, red-coloured granite. The rugged scenery in this area can be appreciated from a drive along Rhino Valley, where the hills drop steeply into the Tsavo River on their southern side. A small inselberg known as Poacher's Lookout, which consists of mica schist has views over the regional plateau to the snow-clad peak of Kibo (Plate 14.4a). The rocks of the Mozambique Belt are typically associated with poor, sandy soils that support scrubby bush and short grasses, distinct from the more thickly vegetated Pleistocene lava flows.

The Mozambique Belt contains significant deposits of gemstones, including sapphirine, hibonite and ruby as well as the locally named *tsavorite* (Sect. 4.7). Small deposits of graphite may be associated with some of the gemstone localities, whilst in some localities in the Chyulu Hills gemstones occur in xenoliths i.e. fragments of Mozambique Belt that were stoped off and transported upward during emplacement of the much younger, rift-related volcanics (Ulianov and Kalt 2005).

14.5.2 Pleistocene and Holocene Volcanism

The oldest lava flows within the park are the Mzima Lavas. They originated in the Chyulu Hills and flowed southeast towards the Tsavo River. The Mzima Lavas have a Late Pleistocene age of 26,000 BP. They can be observed as low, scrub-covered ridges and hill slopes in the vicinity of the Mzima Springs, an area where they have been partially overlapped by younger activity, including Holocene age flows. The most recent volcanism in this area is associated with the Shaitani and Chaimu events, which are thought to have occurred in 1865–1866. Lava flows associated with these events are mostly unvegetated and can be observed cutting prominent black swathes through the verdant bush-covered, older flows. The Chaimu event includes cin-der and ash cones as well as several prominent lava flows (Plate 14.4a). A track accesses the highest cone and provides good views of the associated lava flows. A unique feature of this area is the opportunity to examine an unconformity where granite of the Mozambique Belt has been overlain by the Chaimu lavas, a break in the geological record of at least 500 Ma (Plate 14.4b).

The Shaitani lava flow cuts the road between Tsavo West and Oloitokitok, to the southwest of one of the park gates. This flow is dominated by ropy-textured lavas and blocks of cinder. A prominent central ridge with cooling cracks can be observed (Plate 14.1). The flow can be traced to a poorly vegetated cinder cone to the north of the road with a small crater on the southeastern slopes (Plate 14.5a). A lava tube occurs on the lower, eastern slopes of this cinder cone. The lineation of the lava tube can be estimated on the surface, from clumps of trees spaced about 50–100 m apart that are associated with roof collapse. The immature drainage pattern on the younger cones in the vicinity of the Shaitani cone is a notable feature.

14.5.3 Mzima Springs

The primary water source for Tsavo West is the Mzima Springs (Ertuna 1979). The springs feed the Tsavo River with some 280,000 L/min of fresh, clear water that bubbles up over basaltic boulders. Prior to draining into the river, the springs have formed large pools fringed by giant hardwoods and palms (Plate 14.5b). Water is fed from underground sources in the Chyulu Hills over a distance of at least 7 km. A key component of the system that provides the water is the high porosity of the cinder-dominated volcanic sequences, in contrast to the impermeable Basement Complex. The water is filtered by the volcanic rocks and provides a clean resource, not only to wildlife within the park but some 10% is supplied to Mombasa by a gravity-driven pipeline.

Plate 14.2 **a** View from Noomotio Hill, Amboseli, reveals the eroded roots of a parasitic volcanic cone (part of the Kibo volcanism), enclosed by swamps fed by groundwater from the Kilimanjaro massif. The arid plains that are so characteristic of Amboseli are visible in the background; **b** Calcrete terraces on the margin of the dry bed of Lake Amboseli

Plate 14.3 **a** View looking north from the dry plains of Tsavo West along the spine of the Chyulu Hills; **b** Mtito Andei entrance gate to Tsavo West National Park, 'Ancient Land of Lions and Lava'

Plate 14.4 **a** The Chaimu cinder cone, Tsavo West is sparsely vegetated; **b** The Chaimu lava flow (1865–1866) unconformably overlies the approximately 550 Ma old granite of the Mozambique Belt, Tsavo West

Plate 14.5 a The Shaitani lava flow was erupted from a vent on the southeastern side of a cinder cone in the Chyulu Hills; **b** Lush oasis of the Mzima Springs, Tsavo West, which includes an underwater viewing chamber

References

Ertuna, C. (1979). Flow regime of the Mzima Springs in Kenya. *ICE proceedings, 67,* 833–840.

Gregory, J. W. (1896). *The Great Rift Valley* (421 p). London: John Murray.

Haug, G. H., & Strecker, M. R. (1995). Volcano-tectonic evolution of the Chyulu Hills and implications for the regional stress field in Kenya. *Geology, 23,* 165–169.

Mosley, P. P. (1993). Geological evolution of the late Proterozoic "Mozambique Belt" of Kenya. *Tectonophysics, 221,* 223–250.

Saggerson, E. P. (1963). *Geology of the Simba-Kibwezi area* (Report 58, 70 p). Geological Survey of Kenya.

Shackleton, R. M. (1986). Precambrian collision tectonics in Africa. In M. P. Coward & A. C. Ries (Eds.), *Collision tectonics* (Special Publication Vol. 19, pp. 324–349). Geological Society London.

Spath, A., Le Roex, A. P., & Opiyo-Akech, N. (2000). The petrology of the Chyulu Hills volcanic province, southern Kenya. *Journal of African Earth Sciences, 31,* 337–358.

Spath, A., Le Roex, A. P., & Opiyo-Akech, N. (2001). Plume-lithosphere interaction and the origin of continental rift-related alkaline volcanism: The Chyulu Hills volcanic province, southern Kenya. *Journal of Petrology, 42,* 765–787.

Thomson, J. (1887). *Through Masailands: A journey of exploration among snow-clad volcanic mountains and strange tribes of eastern Africa* (364 p). London: Sampson Low, Marston, Searle, & Rivington.

Ulianov, A., & Kalt, A. (2005). Mg-Al sapphire and Ca-Al hibonite bearing granulite xenoliths from the Chyulu Hills volcanic field, Kenya. *Journal of Petrology, 47,* 901–927.

Lakes of the Gregory Rift Valley: Baringo, Bogoria, Nakuru, Elmenteita, Magadi, Manyara and Eyasi

15

Abstract

The lakes of the Gregory Rift Valley are protected in national parks, reserves and conservation areas. Many of the lakes occur in areas of spectacular landforms located at the base of prominent escarpments. Palaeo-lakes were on average far larger (and deeper) during the Pleistocene. Currently, lakes are mostly relatively small, finger-shaped bodies with a maximum length of a few tens of kilometres. Lake basins have limited catchment and few outlets; rift platforms are tilted outwards so that major rivers flow away from the rift valley. Evaporation exceeds inflow resulting in high levels of alkalinity and salinity. The high sodium content is enhanced due to erosion of the sodium-rich volcanic rocks that characterise the Gregory Rift Valley. Lakes Bogoria, Nakuru, Elmenteita, Manyara and Eyasi are typical in this regard: they are finger-shaped, extremely shallow and markedly alkaline. They have an average pH of 10. Lakes Magadi and Natron are extraordinarily toxic; they contain brines with a pH of 12 that fossilises trees and animals by replacing the wood or bones with sodium carbonate. The alkaline lakes include salt deposits which are a mixture of two naturally occurring compounds of sodium carbonate, trona and natron. The salt deposits of Lake Magadi are quarried for soda ash. Lakes Baringo and Naivasha differ from the alkaline lakes in that they have a more rounded form, occur in basins with larger catchments and are dominated by freshwater. Important vertebrate and hominin fossils have been discovered in the Miocene–Pleistocene sediments of the Tugen Hills, near Lake Baringo. Geysers and hot springs discharge sulphurous brines into some lakes, including Baringo and Manyara. The geysers associated with Lake Bogoria are particularly well known. Many lakes and foreshores sustain significant concentrations of wildlife and they are refuges for more than 400 species of birds. Huge concentrations of flamingoes occur on some of the alkaline lakes.

Keywords

Alkaline lakes • Flamingo • Geysers • Natron RAMSAR • Sodic brines • Trona

Photographs not otherwise referenced are by the author.

Plate 15.1 Hot springs and algae associated with geothermal activity on the foreshore of Lake Bogoria. The Siracho Escarpment (the eastern wall of the rift valley) is visible in the background

15.1   Introduction

Many of the lakes in the Gregory Rift occur in areas of spectacular landforms and are protected in national parks, reserves and conservation areas. The larger lakes are restricted to the rift valley where they reveal characteristic finger shapes and are generally shallow, alkaline features. The lake basins are relatively small with few outlets. The restricted catchment means that evaporation typically exceeds inflow. Five of the lakes (from north to south), Bogoria, Nakuru, Elmenteita, Manyara and Eyasi are typical in this regard (Fig. 15.1). They have an average pH of 10. Lakes Magadi and Natron occur in an extremely arid area where extraordinarily toxic brines have developed. Lakes Baringo and Naivasha differ from the alkaline lakes as they report more rounded shapes and are dominated by freshwater. Lake Naivasha (Chap. 16) and Lake Natron (Chap. 17) are described separately, as is the ephemeral Lake Amboseli (Chap. 14). Small lakes occur in some of the large calderas, e.g. Ngorongoro and Empakaai (Chap. 10). The Momella lakes occur in the debris avalanche deposit (DAD) associated with Mount Meru (Chap. 13). Some lakes in the rift valley are associated with geysers and other evidence of geothermal activity (Plate 15.1). Lakes typically have unusually high contents of fluorine, in addition to sodium, both elements derived from weathering of the alkaline volcanic rocks which are so charcteristic of the region (Baker 1987; Dawson 2008). According to Gaciri and Davies (1993), some rural communities suffer from the disease of fluorosis, a result of drinking untreated water high in fluorine and exasperated by a regular diet of lake fish.

15.2   Palaeo-Lakes

The lakes of the East African Rift System (EARS), including those in the Albertine Rift (Chap. 3) provide a unique record of climatic changes over the past several million years (Nicholson 1996). Street and Grove (1979) have focused on changes over the previous 30,000 years with data from fluctuations in the size and depth of palaeo-lakes. Some of these changes reflect climatic shifts on a global scale. The palaeo-lakes of the Gregory Rift Valley varied considerably as a consequence of climatic cycles during the Late Pleistocene and Early-mid Holocene (Olago et al. 2009). The size and depth of lakes have an important effect on the alkalinity and salinity. During the wetter periods of the Holocene (Sect. 6.3), many lakes that are now shallow and alkaline were far deeper and freshwater. Some lakes were conjoined into much larger palaeo-lakes (Fig. 15.1) and many of the lake basins were reshaped during the tectonically active Late Pleistocene (and to a lesser extent during the Early-mid Holocene). The differential rise of rift platforms in some areas has had the effect of channelling water into distal basins whilst bypassing nearby lakes. Finally, the rise and tilting of rift platforms has persistently reduced the overall flow of rivers into the rift valley.

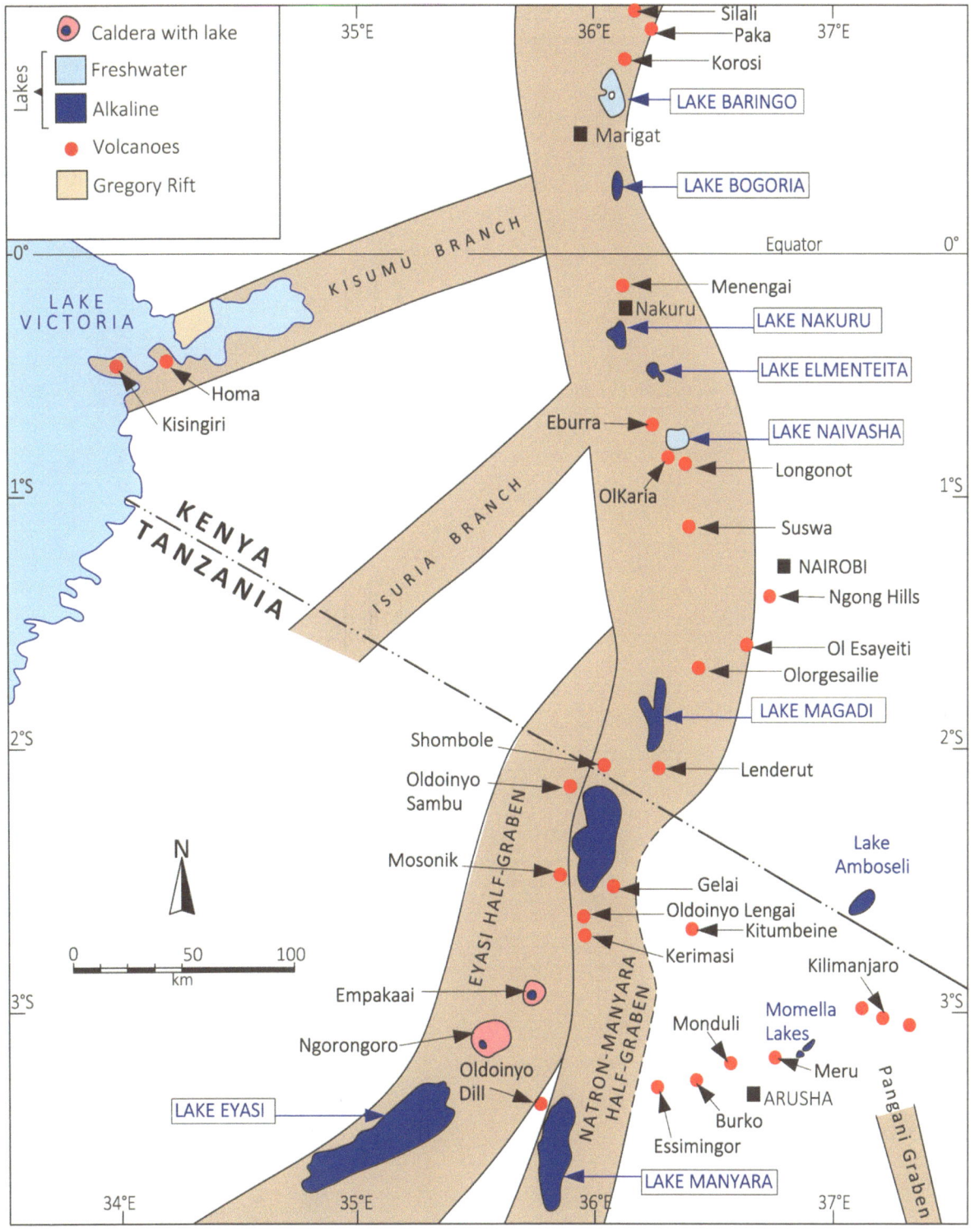

Fig. 15.1 Lakes in the Gregory Rift Valley of central/southern Kenya and northern Tanzania. Small lakes occurring outside of the rift valley are restricted to very specific localities such as calderas (e.g. Ngorongoro), debris avalanche deposits (e.g. Momella), or warps on the regional plateaus (e.g. Lake Amboseli)

15.3 Lake Baringo National Park

Lake Baringo, the most northerly of the lakes described here, has a distinctive rounded shape and a surface area of 130 km². The lake and its islands are protected in a national park served by the regional town of Marigat. Lake Baringo occurs in a relatively low-lying (970 m) section of the Gregory Rift Valley which includes Lake Bogoria (Fig. 15.2). The rift here is comprised of a half-graben with prominent escarpments restricted to the eastern side. The

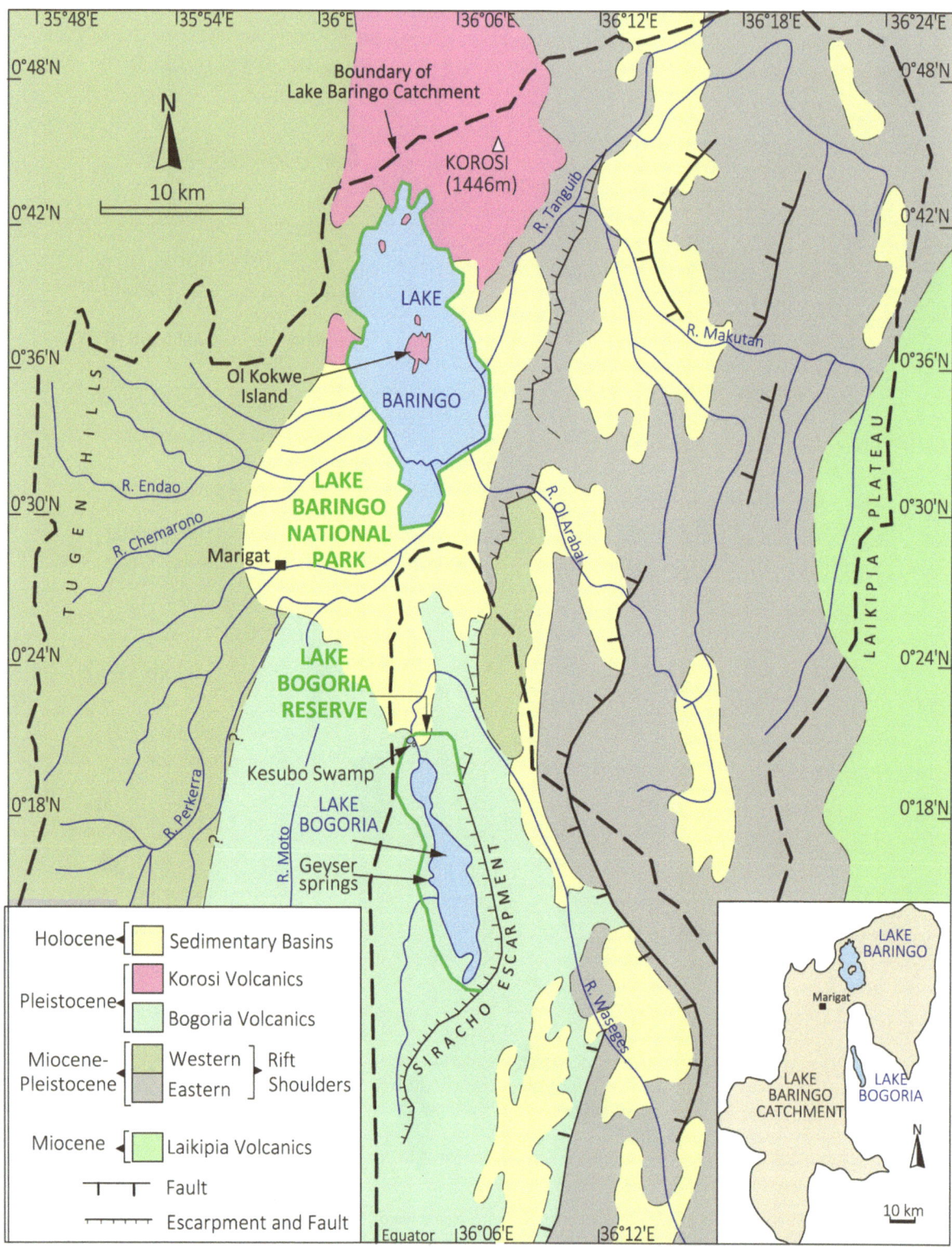

Fig. 15.2 Geological map of the area around Lakes Baringo and Bogoria simplified from various sources including Woolley (2001)

western side of the valley is rather poorly defined. Lake Baringo is fed from a relatively large basin with rivers flowing from both the north and south. The absence of outlets causes the depth of the lake (average of 5 m) to vary considerably. The level has risen almost 5 m in recent years, probably due to increased rainfall. Lake Baringo supports fishing for both the local population and tourists (several endemic freshwater fish occur). During times of drought, when grazing was poor on the foreshore, one of the local tribal groups, the Njemps (or Il Chamus) also known as the 'People with the Swimming Cows' traditionally migrated to the islands with their cattle swimming next to their canoes.

The unusual shape of the Baringo catchment (inset, Fig. 15.2) is ascribed to detachment blocks on the edge of the Gregory Rift Valley, i.e. the Aberdare Mountains to the east and the Elgeyo Block to the west. Two examples illustrate the complexity of the drainage due to recent tectonism. The Molo River drains part of the Mau Escarpment near Lake Natron and yet flows northwards, skirting Lake Bogoria before feeding Lake Baringo. Conversely, some rivers that rise near Lake Baringo channel water either northwards into Lake Turkana or southwards into Lake Nakuru. The uplifted fault block on the eastern extremity of the Baringo basin is comprised of the Early Miocene-age Laikipia Plateau Volcanics (Woolley 2001). The shoulders of the eastern and western rift platforms are dominated by volcanic groups which cover Late Miocene, Pliocene and Pleistocene ages. The volcanics include thick units of resistant lavas through which rivers have carved narrow gorges (Plate 15.2a).

The stratigraphy of the Baringo basin has been investigated using radiometric dating of the volcanic rocks as part of calibrating the interrelated fossiliferous sedimentary units in the region. These latter include the Tugen Hills, an uplifted block located in this relatively wide section of the rift valley, which encompass a period of at least 14 Ma, from the Late Miocene through to the Pleistocene. Paleontological finds include vertebrates which have yielded evidence of a relatively rapid evolutionary pattern (Hill et al. 1986). The Tugen Hills has also yielded important hominin finds (Wood 1999). Fossils from this area are exhibited in the Smithsonian Museum, Washington.

The centre of the rift valley in the vicinity of Lake Baringo is dominated by Pleistocene-age volcanics (Fig. 15.2). They include the Korosi Volcanics (0.38–0.10 Ma), derived from a large cone that rises some 500 m above the valley floor to the northeast of the lake (Woolley 2001). The dominantly basaltic and trachytic lavas of this volcano cover an area of 260 km^2, including large areas around the lake and several islands. The northern part of the Ol Kokwe Island is formed by a coalescence of basaltic scoria cones (Plate 15.2b), with the southern part associated with trachytic lava flows. Sulphur-rich fumaroles, hot springs and gossans (hydrothermally altered rocks with oxidised sulphides) occur on this and other islands in Lake Baringo (Fig. 15.3a). Discharge into the lake causes small bays to concentrate sulphurous waters. A large sedimentary basin dominated by Holocene fill occurs to the south of Lake Baringo.

15.4 Lake Bogoria National Reserve

J. W. Gregory described Lake Bogoria, where the Siracho Escarpment rises some 600 m above the rift valley as having the 'most beautiful view in Africa'. This lake is well known for geothermal features including geysers and hot springs (Plate 15.1). The abundance of geothermal features has resulted in sulphurous waters ponding near the foreshore in some areas (Plate 15.3a). The reserve was established in 1973 to protect the remote wilderness area of an area formerly known as Lake Hannington (Plate 15.3b). The northern foreshore can be a relative hostile environment with extensive salt flats and volcanic cliffs (Plate 15.3c). The length of 35 km and width of 3.5 km is characteristic of many of the alkaline lakes of the Gregory Rift Valley, as is the alkalinity (pH of 10.5). Most water in the Bogoria basin is sourced from the Waseges (or Sandai) River, which rises on the lower slopes of the Aberdare Range (Fig. 15.2). Lake Bogoria is somewhat deeper (maximum depth of 10 m) than many of the other alkaline lakes. The absence of outlets has retained a palaeo-stratification: less dense surface waters overlie denser, more saline basal waters. The composition shows considerable variability and high concentrations of the metals sodium, potassium, arsenic, fluorine and the rare earth elements, typical of the surrounding volcanic rocks have been identified (Jirsa et al. 2013). Several hundred hot (alkaline) springs located near the lake supply high concentrations of acids (including HCO_3^- and CO_3^{2-}), as discussed by Renaut and Tiercelin (1993, 1994). The rocky, barren shores of the lake include flows of phonolite, part of the Pleistocene-age Lake Bogoria Volcanics (Plate 15.3c). Fluviolacustrine sediments with reddish coloured beds, rich in zeolitic clays, which are up to 1 m thick, occur on the northern foreshore (Renault 1993).

Lake Bogoria was part of an enlarged palaeo-lake which included Lake Baringo in the Early Holocene (Fig. 6.1). Drill cores of sediments underlying the lake reveal evidence of regular changes from alkaline to freshwater during the Late Holocene. Some access roads within the reserve have been submerged since 2013 and the level remains anomalously high. New access tracks, cut through the poorly vegetated lava flows, enhance the feeling of a barren landscape. Geothermal activity on the shores of the lake includes geysers, hot springs and steam jets (Renaut and Owen 2005). Eighteen geysers have been documented (they once spouted to heights of 5 m), although activity has waned in recent years and is currently restricted to small fountains and hot pools, typically with abundant sulphur and algae. The colour of the hot pools is related to temperature, a feature that is well known from the geyser fields in the Yellowstone National Park in the state of Wyoming. Yellow and green pools contain algae and report temperatures of 60–70 °C and 50–60 °C, respectively. The brown pools contain mosses, crustaceans and insects at temperatures of 27–50 °C.

15.5 Lake Nakuru National Park

Lake Nakuru is located near the regional town of the same name and has long been protected in a national park famous for both large herbivores and aquatic birds (Fig. 15.3). The lake occurs at an altitude of 1,754 m, considerably higher than the lakes farther north. The size is very variable (5–45 km^2) and the depth is rarely more than 3 m. In the early 1990s, the lake almost dried up entirely. The lake is moderately alkaline (pH of 10). During a drought in 2009 when the water level dropped drastically, the total salinity rose from 20% to a remarkable 63% (Jirsa et al. 2013). In recent years, however, the lake has been so full as to be less saline and has flooded parts of the foreshore (Plate 15.4a). The principal water source is the Njoro River which flows eastwards from the Mau Forest. Aquifers in the basin are composed of lacustrine deposits derived from erosion and redeposition of volcanic rocks (McCall 1967). Views from Baboon Cliff, an eroded lava flow on the western shore provide an excellent panorama (Plate 15.4b). Hyrax Hill is an important archaeological site with Neolithic and Stone Age remains.

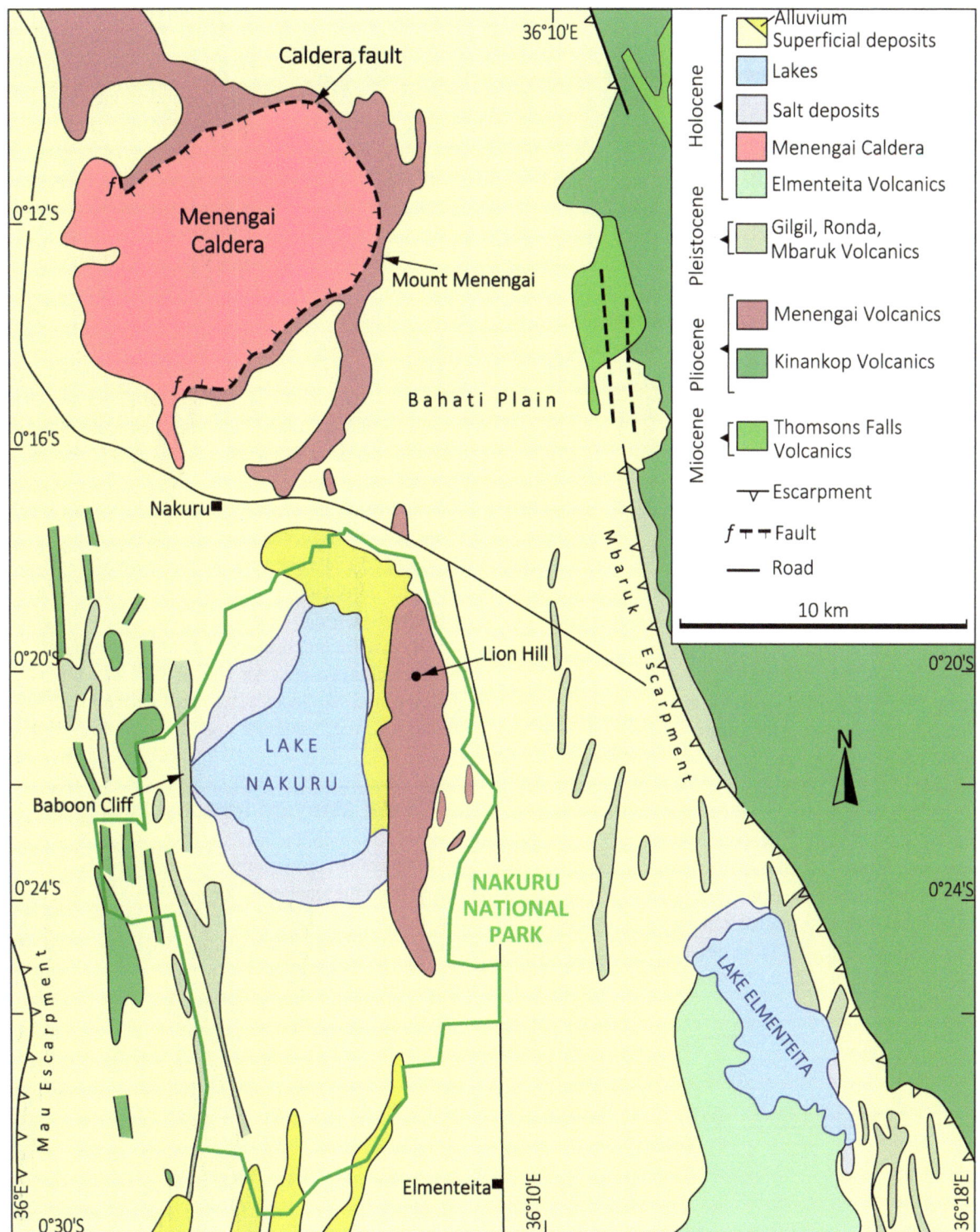

Fig. 15.3 Geological map of the area around Lakes Nakuru and Elmenteita simplified from the Nakuru Quarter Degree Sheet 43 (1966) and accompanying report by McCall (1967)

15.6 Lake Elmenteita

Lake Elmenteita is located close to the Mbaruk Escarpment to the south of Lake Nakuru (Fig. 15.3). The lake is extremely shallow (less than 1 m deep) and moderately alkaline (average pH of 10). Lake Elmenteita is fed from the Mereroni River, which drains the Bahati Plain and highlands to the north and east within a small basin. Elmenteita may be connected with Lake Nakuru via underground aquifers. The lake is a protected RAMSAR site due to its birdlife and has been named by UNESCO as a world heritage site (Plate 1.2 e, f). The Kekopey hot springs at the southern end of the lake are the breeding site for an introduced species of tilapia that may have caused the flamingo population to decrease markedly. It has been estimated that the tilapia caused over a million flamingoes, which formerly bred on Lake Elmenteita, to migrate to Lake Natron.

15.7 Lake Magadi

Lake Magadi occurs in a large basin located at a relatively low altitude of approximately 600 m in southern Kenya. This section of the Gregory Rift Valley is unusually desolate and reports a hot, arid climate (Baker 1958; Vincens and Casanova 1987). The Magadi basin is surrounded by volcanic cones and plateaus with altitudes of up to 3,000 m. Views include the extensive Nguruman Escarpment on the western side of the rift (Plate 15.5a). The lake is rarely more than 1 m deep and in dry seasons, approximately 80% is covered by salt deposits. The concentrated and extremely dense alkaline brines (pH of 12) are caustic enough to burn the skin. Small pools on the margins of the lake are even more toxic and tree and animal remains are fossilised by sodium carbonate. There is little surface run-off and saline springs with temperatures of up to 85 °C are the main supply of brines. Lake Magadi contains a single species of cichlid fish (*Alcolapia grahami*) that has evolved to cope with both the extreme alkalinity and high temperatures (up to 45°). Extensive salt deposits have precipitated from the sodic brines in the vicinity of the lake (Eugster 1970, 1980; Jones et al. 1977). They include layers of sodium carbonate salt up to 40 m thick, some of which are quarried to produce soda ash. The salt is a mixture of two minerals, *natron* (mix of sodium carbonate decahydrate [$Na_2CO_3.10H_2O$] and sodium bicarbonate [$NaHCO_3$]) and *trona* (sodium sesquicarbonate dihydrate [$Na_3CO_3.HCO_3.2H_2O$]) (Behr 2002). The salt deposits are intercalated with the *Magadi chert* (microcrystalline silica) which is derived from the dissolution of a rare sodium silicate mineral (*magadiite*) first identified here. The chert also occurs as dyke-like bodies that were injected into the bedded deposits when the silica was soft enough to be deformed.

The source of the water in the Magadi–Natron Basin was investigated by Kaufmann et al. (1990) using the [36]Cl methodology. The chemistry of the surface brines is consistent with derivation from precipitation (with no evidence of dissolved Cl entering the system). Deeper level brines, however, reveal compositions indicative of recharge during a different climatic era. An age of 0.76 Ma for the accumulation of the salt deposits has been determined. This may correlate with the maximum age of the lake.

15.8 Lake Manyara National Park

Lake Manyara occupies some two-thirds of the park that is accessed from the small town of Mto-Wa-Mbu which nestles below the Western Escarpment of the Gregory Rift Valley, northern Tanzania. The setting, described by Ernest

Hemingway as 'the loveliest lake in Africa' includes views of the steep, thickly forested escarpment (Plate 5.1). The eastern side of the valley in this area (a half-graben) is a gently sloping plain. The extent (maximum area of 230 km^2), depth (average of 3.7 m) and alkalinity (pH of 9.5) of Lake Manyara vary considerably. Large areas of salt flats and mud banks are exposed during dry seasons. The lake is fed by the Simba and Makayuni Rivers. Hot springs and pools with multicoloured algae occur near the south-western shore (Plate 15.5b). The basin is partially infilled by the Manyara Group, a thick sequence of sediments and volcaniclastics (derived from the Ngorongoro Highlands) that have been described and investigated in detail by Schwartz et al. (2012). Modern vertebrate tracks have been used to ascertain the palaeobiological setting of the lake (Cohen et al. 1993). The western shore of Lake Manyara is nurtured by underground springs replenished from the Ngorongoro Highlands. The Kwakuchinja wildlife corridor, situated to the east of Lake Manyara, allows wildlife to migrate between the Tarangiri National Park, the Ngorongoro Highlands and the Serengeti Plains (Fig. 1.1).

15.9 Lake Eyasi

The seasonal, highly saline and alkaline Lake Eyasi, the southernmost of the large lakes in the Gregory Rift Valley is located below a huge escarpment capped by volcanic rocks of the Ngorongoro Volcanic complex. The principal inflow to the lake is from the Sibiti River on the southern end. The lake may dry completely during the dry season—when it can be exposed to severe aeolian deflation—and even during the wet season depths rarely exceed 1 m. The Mumba Rock Shelter has yielded important archaeological findings that include stone artefacts (Late Middle Stone Age as well ostrich eggshell beads. Fossil evidence suggests that *Homo*

sapiens lived in the approximately 20 m high shelter at 130,000 BP. Lake levels in recent years have been sufficiently high as to partially submerge the shelter.

15.10 RAMSAR Status and Flamingo

Most lakes in the Gregory Rift Valley have been afforded RAMSAR status as they contain important wetlands. Some lakes are sanctuaries for as many as 500 bird species. Lakes Baringo and Naivasha in particular reveal a great diversity of birds, and reserves here include large mammals and reptiles. Concentrations of endangered species of large mammals are most notably protected in the Nakuru and Manyara National Parks. The latter is well known for studies of elephant behaviour. Many of the alkaline lakes are famous for one of the world's greatest spectacles: the concentration of flamingo (Plate 1.2e, f). They were formerly so numerous (several millions), as to change the colour of shorelines, most notably on Lake Nakuru. The flamingo has evolved remarkably efficient filtration systems to cope with lake brines that are poisonous to most other species. Their principal food is cyanobacteria which flourish in highly alkaline waters. At their peak, the flamingos were so numerous on Lake Nakuru, they consumed some 250,000 kg of algae per hectare of surface water annually. Some 20 years ago the number of flamingo decreased, either due to pollution from increased land usage in catchments, or simply higher rainfall. High water levels since 2013 have exacerbated the problem as decreased alkalinity inhibits algal growth. The loss of flamingo from Lake Nakuru was initially balanced by an incremental increase at Lake Bogoria, but recently the numbers have decreased here as well. Whether pollution or non-reversal climatic shifts are the cause is not known. In summary, lakes in the Gregory Rift are extremely sensitive to both climatic fluctuations and land-use changes in their catchment.

Plate 15.2   a Rivers in the Lake Baringo basin have carved narrow gorges in the volcanic lavas on the western side of the rift valley; **b** Basaltic lavas on Ol Kokwe Island, Lake Baringo

Plate 15.3 **a** Sulphurous waters near geysers and hot springs pond on the edge of Lake Bogoria; **b** The Lake Bogoria National Reserve was established in 1973 to conserve the area around 'Lake Hannington'; **c** The northern foreshore of Lake Bogoria includes extensive salt flats with cliff faces (foreground) of the Bogoria Volcanics and the Siracho Escarpment (background)

Plate 15.4 **a** The high water level in 2015 resulted in flooding of the northern foreshore of Lake Nakuru; **b** View of Lake Nakuru from Baboon Cliff, a resistant rampart of lavas and ashes

Plate 15.5   **a** View of the soda ash works on the eastern shore of Lake Magadi (centre) with part of the Nguruman Escarpment (background) in southern Kenya. Source: Public domain website https://www.flickr. com/photos/ninara/15563504833; **b** Hot springs have created deposits of travertine with algae on the western shore of Lake Manyara

References

Baker, B. H. (1958). *Geology of the Magadi area* (Report 42, 81 p). Geological Survey of Kenya.

Baker, B. H. (1987). Outline of the petrology of the Kenyan Rift alkaline province. In J. G. Fitton & B. G. J. Upton (Eds.), *Alkaline igneous rocks* (Vol. 30, pp. 293–311). London: Geological Society of London Special Publication.

Behr, H. J. (2002). Magadiite and Magadi Chert: A critical analysis of the silica sediments in the Lake Magadi Basin, Kenya. *SEPM Special Publication, 73*, 257–273.

Cohen, A. S., Halfpenny, J., Lockley, M., & Michel, E. (1993). Modern vertebrate tracks from Lake Manyara, Tanzania and their paleobiological implications. *Paleobiology, 19*(4), 433–458.

Dawson, J. B. (2008). *The Gregory Rift Valley and neogene-recent volcanoes of northern Tanzania* (Vol 33, 102 p). Geological Society London Memoirs.

Eugster, H. P. (1970). Chemistry and origin of the brines from Lake Magadi, Kenya. *Mineralogical Society of America Special Paper, 3*, 215–235.

Eugster, H. P. (1980). Lake Magadi, Kenya, and its Pleistocene precursors. In A. Nissenbaum (Ed.), *Hypersaline brines and evaporitic environments* (pp. 195–232). Amsterdam: Elsevier.

Gaciri, S. J., & Davies, T. C. (1993). The occurrence and geochemistry of fluoride in some natural waters of Kenya. *Journal of Hydrology, 143*, 393–412.

Hill, A., Curtis, G., & Drake, R. (1986). Sedimentary stratigraphy of the Tugen Hills, Baringo, Kenya. *Special Publication Geological Society London, 25*, 285–295.

Jirsa, F., Gruber, M., Stojanovic, A., Omondi, S. O., Mader, D., Korner, W., et al. (2013). Major and trace element geochemistry of Lake Bogoria and Lake Nakuru, Kenya, during extreme drought. *Chem Erde, 73*(3), 275–282.

Jones, B. F., Eugster, H. P., & Rettig, S. L. (1977). Hydrochemistry of the Lake Magadi basin, Kenya. *Geochimica et Cosmochimica Acta, 41*, 53–72.

Kaufmann, A., Mordeckai, M., Paul, M., Hillaire-Marcel, C., Hollos, G., Boaretto, E., et al. (1990). The ^{36}Cl ages of the brines in the Magadi-Natron basin, East Africa. *Geochimica et Cosmochimica Acta, 54*, 2827–2833.

McCall, G. J. H. (1967). Geology of the Nakuru-Thomson's Falls-Lake Hannington area. *Explanation to degree sheet 35 (SW quadrant) and 43 (NW quadrant)* (Report 78, 122 p). Geological Survey of Kenya.

Nicholson, S. E. (1996). A review of climate dynamics and climate variability in Eastern Africa. In T. C. Johnson & E.O. Odada (Eds.), *The Limnology, Climatology and Paleoclimatology of the East Africa Lakes* (pp. 25–56). Amsterdam: Gordon and Breach.

Olago, D., Opere, A., & Barongo, J. (2009). Holocene palaeohydrology, groundwater and climate change in the lake basins of the Central Kenya Rift. *Hydrological Sciences Journal, 54*(4), 765–780.

Renault, R. W. (1993). Zeolitic diagenesis of late Quaternary fluvio-lacustrine sediments and associated calcrete formation in the Lake Bogoria Basin, Kenya Rift Valley. *Sedimentology, 40*(2), 271–301.

Renaut, R. W., & Tiercelin, J.-J. (1993). Lake Bogoria, Kenya: Soda, hot springs and about a million flamingos. *Geology Today, 9*, 56–61.

Renaut, R. W., & Tiercelin, J.-J. (1994). Lake Bogoria, Kenya Rift Valley: a sedimentological overview. In R. W. Renaut & W. M. Last (Eds.), *Sedimentology and Geochemistry of Modern and Ancient Saline Lakes* (Vol. 50, pp. 101–123). Tulsa: SEPM Special Publication.

Renaut, R. W., & Owen, R. B. (2005). The geysers of Lake Bogoria, Kenya Rift Valley, Africa. *GOSA Transactions, 9*, 4–18.

Schwartz, H., Renne, P. R., Morgan, L. E., Wildgoose, M. M., Lippert, P. C., Frost, S. R., et al. (2012). Geochronology of the Manyara Beds, northern Tanzania: New tephrostratigraphy, magnetostratigraphy and ^{40}Ar/^{39}Ar ages. *Quaternary Geochronology, 7*, 48–66.

Street, F. A., & Grove, A. T. (1979). Global maps of lake fluctuations since 30,000 yr BP. *Quaternary Research, 12*, 83–118.

Vincens, A., & Casanova, J. (1987). Modern background of Natron-Magadi basin (Tanzania-Kenya): Physiography, climate hydrology and vegetation. *Science Geology Bulletin (Frankfurt), 40*(1–2), 9–21.

Wood, B. (1999). Plio-Pleistocene hominins from the Baringo Region, Kenya. In P. Andrews & P. Banham (Eds.), *Late cenozoic environments and hominid evolution: A tribute to Bill Bishop* (pp. 113–122). Geological Society of London.

Woolley, A. (2001). *Alkaline rocks and carbonatites of the world. Part 3: Africa* (372p). Geological Society of London.

Lake Naivasha and the Mount Longonot and Hell's Gate National Parks

Abstract

Lake Naivasha, is the highest of the rift valley lakes in Kenya, occurring at an altitude of 1,884 m. The lake is rimmed by substantial swamps and the average depth of 6 m is highly variable. Two active volcanic systems, Mount Longonot and the Olkaria Volcanic complex occur on the southern shores. The Longonot Volcano includes a prominent cone (2,776 m) with a well-defined summit crater; the most recent eruption was in 1860. The cone is located within a much larger caldera; multiple caldera events triggered huge outpourings of lavas and ignimbrites in the period 21,000–6,000 BP. The volcanism at Longonot is dominated by trachytic basalt lavas with abundant pyroclastic deposits. The geology of the Hell's Gate National Park, named from the lunar landscapes in a valley fringed by lava cliffs, is dominated by sections of the Olkaria Volcanic complex. This last erupted approximately two hundred years ago. The Njorowa Gorge, a narrow slot rimmed in by near-vertical sidewalls, exposes a section of finely bedded ashes which accumulated from explosive eruptions of rhyolitic magmas. The gorge formed when the much larger, palaeo-Lake Naivasha drained southwards during the Late Pleistocene–Early Holocene. Many of the eruptions at Olkaria occurred beneath this palaeo-lake. Olkaria is an important source of obsidian or volcanic glass, used by ancient cultures for stone tools. The Mount Longonot Volcano and the Olkaria Volcanic complex are both potentially hazardous and should be carefully monitored, particularly, as they occur proximal to the relatively densely populated area around Lake Naivasha. The style of volcanism at Olkaria can be compared with the Yellowstone Volcano in Wyoming, potentially one of the most destructive volcanoes on Earth. The shallow magma chamber at Olkaria is a major source of geothermal energy that provides much of Kenya's electricity.

Keywords

Cone • Crater • Ignimbrite • Njorowa gorge Obsidian • Rhyolite • Yellowstone

Photographs not otherwise referenced are by the author.

Plate 16.1 The near-vertical walls of Njorowa Gorge, Hell's Gate National Park are composed of multiple layers of finely bedded volcanic ash. The yellow coloration is typical of rhyolitic compositions associated with highly explosive volcanism

16.1 Introduction

The Mount Longonot and Hell's Gate National Parks protect active volcanic centres located on the southern shores of Lake Naivasha, one of the only two freshwater lakes in the Gregory Rift Valley. The proximity of this section of the rift valley to Nairobi results in relatively large numbers of tourists visiting, many of whom hike trails through the volcanic terranes which constitute the main features of this area (Plate 16.1). The two volcanic areas were incorporated into national parks in 1984, with Hell's Gate being awarded UNESCO Heritage status in 2010. Both parks are located within the rift valley (Fig. 16.1a). They provide important migration corridors for large mammals and also host several endangered species. The Mount Longonot Volcano includes an intact cone, characteristic of recent systems that have undergone minimal erosion. Longonot last erupted in 1860. The Hell's Gate National Park is dominated by the Olkaria

Volcanic complex where the most recent activity occurred approximately two hundred years ago. The geothermal energy associated with the magma chamber underlying Olkaria is exploited for a relatively clean source of electricity. The most important economic activity in the vicinity of Lake Naivasha is the growing and export of cut flowers.

16.2 Regional Geology

The extent of the Naivasha basin can be ascertained from the regional geological map (Fig. 5.2). This area of the Gregory Rift Valley is dominated by extensive Pleistocene–Holocene-age volcanic terranes. Three active volcanic systems are identified, Mount Longonot, the Olkaria Volcanic complex and, to the northwest of Lake Naivasha, the Eburru Volcanic complex (Woolley 2001). A simplified geological map shows the principal features of the Longonot Volcano

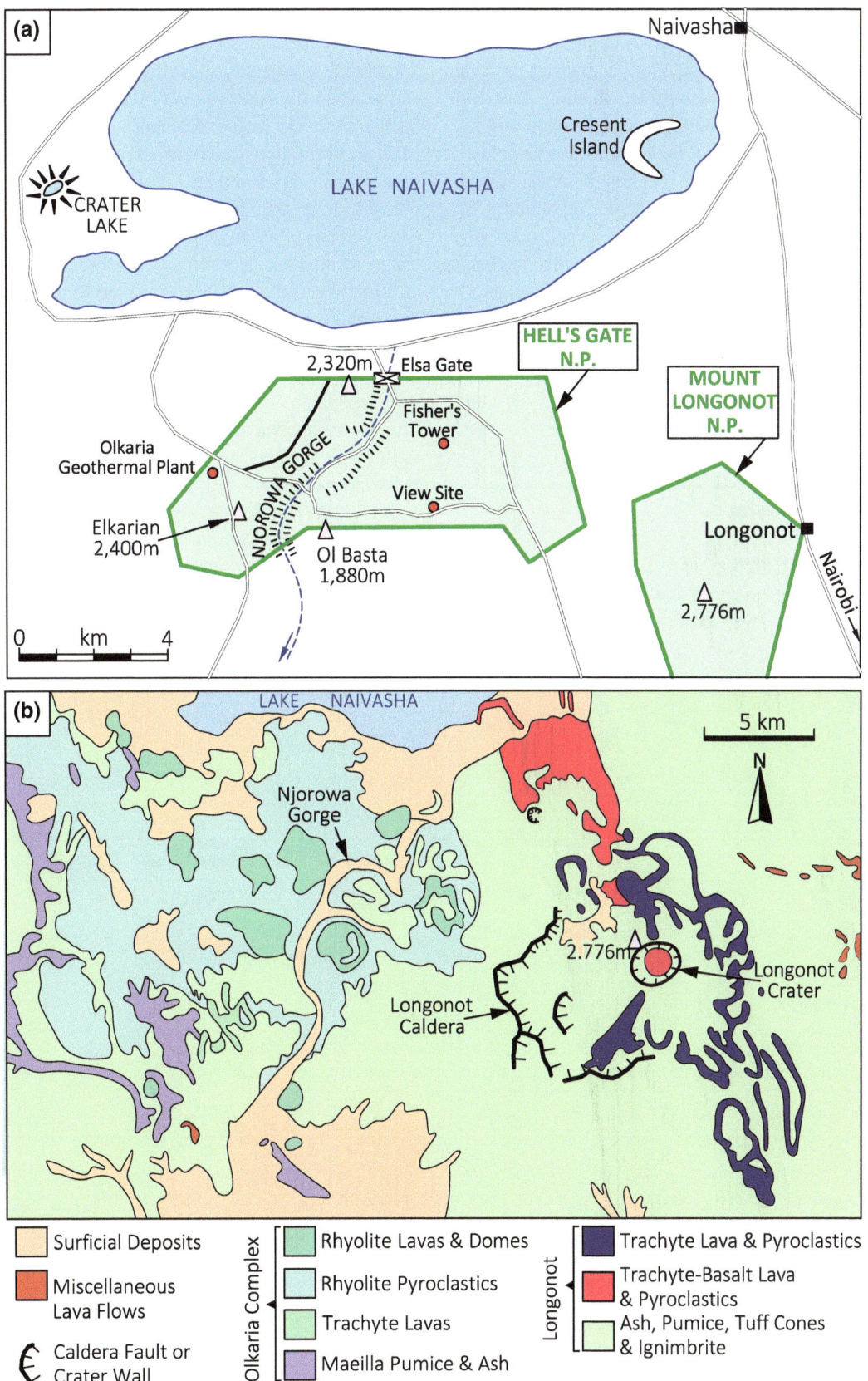

Fig. 16.1 a Location map of Lake Naivasha and Mount Longonot and Hell's Gate National Parks, Kenya; **b** Geological map of the Mount Longonot Volcano and Olkaria Volcanic complex, simplified from Clarke et al. (1990) and Woolley (2001)

and the Olkaria Volcanic complex (Fig. 16.1b). The Olkaria complex is partially obscured by volcanic ash derived from more recent eruptions of Longonot. Numerous volcanic features can be observed in these parks, including the central cone and summit crater of the Longonot Volcano and rhyolitic ashes in the narrow slot of the Njorowa Gorge at Hell's Gate. The volcanology of the Olkaria complex can be compared with the Yellowstone Volcano in Wyoming, USA. Both of these volcanoes are dominated by siliceous (rhyolitic) magmas which generally result in highly explosive eruptions. Yellowstone is described as a super volcano as potential eruption could be catastrophic for large parts of the USA. The Mount Longonot Volcano is also potentially hazardous, and historical Plinian-style eruptions would have had devastating effects on this section of the rift valley during the Late Pleistocene and Early Holocene.

16.3   Lake Naivasha

Lake Naivasha is the highest lake (1,884 m) in the Gregory Rift Valley. The rounded shape and freshwater contrast with the more elongate and alkaline lakes which are so characteristic of the region. The surface area (average of 140 km^2) is periodically enlarged as the lake is rimmed by substantial swamps. The average depth of 6 m is variable, and in 1945 the lake was only 0.6 m deep. The lake is fed from a relatively large catchment that includes the perennial Malewa and Gilgil Rivers. There are no outlets. The Naivasha basin is dominated by Pleistocene–Holocene-age volcanic terranes with subordinate sedimentary rocks. Crescent Island is an exposed rim of an extinct volcanic crater, located within an area of the lake that may be as deep as 30 m. Two small alkaline lakes occur to the southwest of the main lake, Lake Sonachi, which infills a volcanic crater, and Lake Oloiden, the latter often hosting flocks of flamingo.

During the Late Pleistocene and Early Holocene, an enlarged palaeo-lake developed with connections to Lakes Elmenteita and Nakuru (Fig. 15.1). This palaeo-lake drained southwards at Naivasha via the Njorowa Gorge, a narrow slot in the floor of the rift valley. Despite not being protected in a reserve, Lake Naivasha is well known for supporting herds of large game on the foreshore, as well as numerous species of aquatic birds. The growing of cut flowers on the southern shore of the lake is assisted by the nutrient-rich volcanic soils that extend from the northern slopes of Mount Longonot.

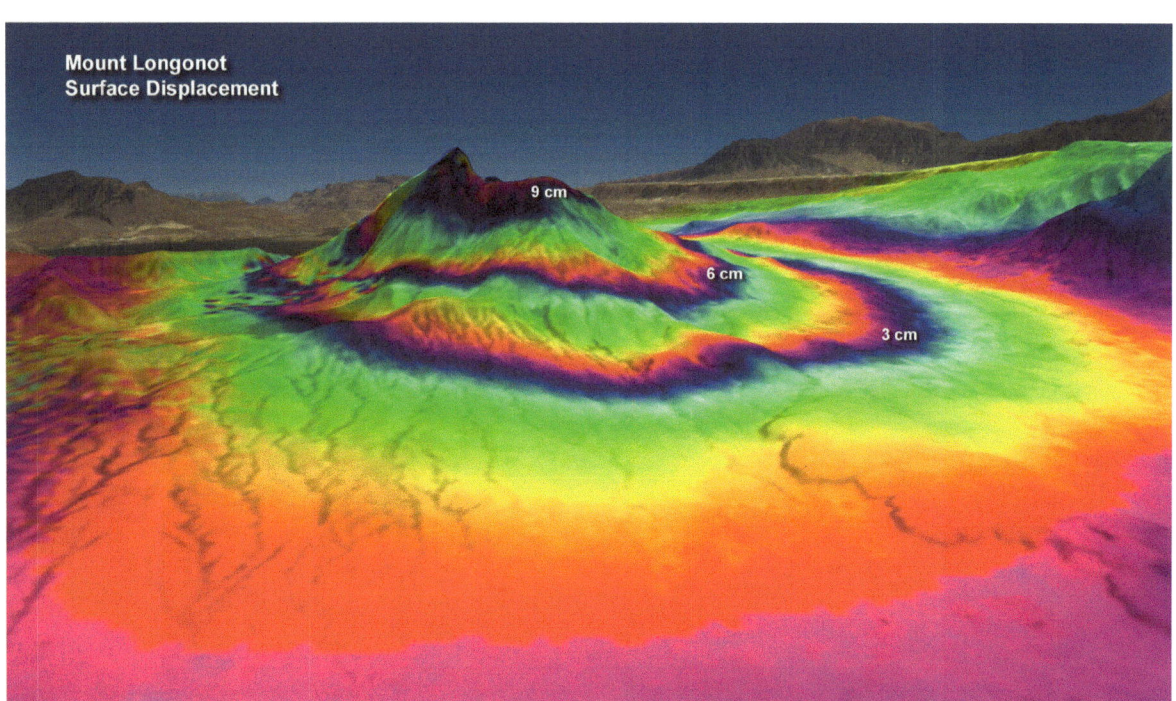

Fig. 16.2 Image showing surface displacement at Mount Longonot over the period June 2004–May 2006 looking northwest over Lake Naivasha towards the Mau Escarpment. The uplift, reported in intervals of 3, 6, and 9 cm, is ascribed to expansion of the subsurface magma chamber. *Source* Interferometric radar (InSAR) elevation observations from ESA's Envisat overlaid on NASA Landsat 7 ETM + image mosaic, courtesy of Planetary Visions/University of Bristol/NERC-COMET/ESA

16.4 Mount Longonot National Park

This relatively small national park protects part of the Mount Longonot Volcano. The volcano covers an area of at least 3,500 km^2 and is far larger than the central cone may suggest (Scott 1980) (Fig. 16.1b). The name 'Longonot' is derived from the Maasai language and describes a mountain with many steep ridges and gullies. The ridges correlate with recent lava flows that have spread down the lower slopes of the cone. The central, relatively intact cone has an elevation of 2,776 m and rises some 800 m above the valley floor (Plate 16.2a). The cone is capped by a near-circular summit crater with a diameter of 1.8 km and steep internal walls, some 350 m in height (Plate 16.2b). The floor of the crater is thickly forested and appears from the rim as an undiscovered wilderness. A prominent parasitic crater occurs on the northern flanks of the cone. An important feature that is not readily apparent to the casual visitor is the presence of a much larger caldera. The caldera, which has been mapped by Scott (1980) and Clarke et al. (1990) measures approximately 12 km by 8 km in diameter. The southern rim correlates with the rugged topography situated several kilometres from the base of the cone (Plate 16.3a). Ash fall associated with the Longonot Volcano extends over an area of as much as 30,000 km^2, covering parts of the Olkaria Volcanic complex and even extending northwards onto the Eastern Rift Platform.

16.4.1 Volcanic Activity

The activity at Longonot can be generalised into four discrete groups or events: (i) early activity; (ii) repeated caldera events; (iii) building of the central cone and summit crater as a consequence of rejuvenation of activity within the older calderas; and (iv) recent lava flows from the summit crater (Clarke et al. 1990). The earliest activity is estimated to have commenced during the Pleistocene at approximately 0.4 Ma. Multiple caldera events which occurred in the period 21,000–6,000 BP triggered huge outpourings of lavas and ignimbrites. Surface exposures of the youngest of the calderas have obscured the older calderas (evidence for this is supported by geophysical surveys). The central cone and summit

crater formed during the mid-Holocene at approximately 3,500 BP. Recent lava flows associated with the summit crater occurred internally and on the flanks of the cone. The fronts to some of the flank flows may be 40 m in height. The youngest of the lava flows can be discerned as they are poorly vegetated. The composition of the Longonot Volcano is dominated by trachytic basalt lavas with abundant pyroclastic deposits (Plate 16.3b), typical of the group (ii) magmas identified by Baker (1987), as described in Sect. 5.9. The eruptive history, however, coincides with the full graben style of rifting in southern Kenya, not the half-graben stage. The bulk of the lavas and ashes at Longonot are not represented in the exposed material, so that calculating average compositions is problematic (Scott 1980).

16.4.2 Magma Chamber

On the basis of detailed studies of tephra from the Longonot Volcano and the adjacent, considerably larger Suswa Volcano (Fig. 5.2), Scott and Skilling (1999) suggested that calderas at both localities formed not from depletion of individual chambers but from a large regional event. All of the volcanic centres in the vicinity of Lake Naivasha, i.e. Eburru, Longonot, Olkaria and Suswa, can be linked to the upwelling of a giant, yet relatively shallow magma chamber. This magma chamber caused a gentle dome structure to develop in the floor of the rift valley. This feature is cut by a series of NNW-SSE lineaments, indicative of extension of the rift. Extension is estimated to have occurred at the rate of 3.2 mm/annum. The caldera events at Longonot and Suswa may have been driven by rapid decompression of this magma chamber in response to this extension. These hypotheses are in part supported by a geodetic survey. The active magma chamber underlying Longonot occurs at a relatively shallow depth and causes the surface to bulge upwards (Fig. 16.2). This has also resulted in the occurrence of near-surface geothermal activity, high on the flanks of the cone and in the summit crater. The absence of a significant geothermal field at Longonot, similar to those associated with the Olkaria Volcanic complex or the Menengai Volcano, may be ascribed to the relative thickness of the young pile of pyroclastics.

16.5   Hell's Gate National Park

The Hell's Gate National Park covers an area of 68 km^2 and occurs at an average elevation of 1,900 m on the southern shores of Lake Naivasha. The park contains abundant game and the sight of plains animals grazing in a barren volcanic landscape is unusual, particularly in the lower parts of the Njorowa Valley. The dominant geological feature is the multicentred Olkaria Volcanic complex, which in its entirety covers an area of 240 km^2 (Fig. 16.1b). More than 80 volcanic centres have been identified at Olkaria. The volcanism in this area commenced in the Late Pleistocene at approximately 45,000 BP (Clarke et al. 1990; Woolley 2001). The complex is underlain by a shallow magma chamber which is manifested by numerous fumaroles and other geothermal activity. Olkaria is comprised of lavas and pyroclastics with a broad range of compositions, including basalt, trachyte and peralkaline rhyolite. They are typical of the group (iii) magmas defined by Baker (1987). The lavas generally crop out as small, steep-sided domes and short flows. The largest dome, Olkaria Hill has a diameter of 2 km and height of 400 m. The pyroclastics constitute laterally persistent sheets which can be observed, for example, in the volcanic cliffs which fringe the lower parts of the Njorowa Valley near the Elsa Gate.

16.5.1   Volcanic Pile

Drilling for the geothermal resource at Olkaria has revealed that the volcanic pile persists to a depth of at least 2,500 m. The oldest component exposed on the surface is the Maiella Formation (Fig. 16.1b), a sequence of pumice, ash and lapilli tuff which was produced by major Plinian-style eruptions (Clarke et al. 1990). This formation is overlain by trachyte lavas of the Olkaria Formation. The latter are partially covered by one of the youngest products of the complex, thick layers of rhyolitic pyroclastics which include ignimbrites. The ignimbrite has a pale grey colour and is readily identified by the presence of welded lithic fragments. Ignimbrites are the product of extensive pyroclastic flows,

one of the most devastating products of Plinian-style volcanism. A thick layer of ignimbrite with distinctive columnar jointing occurs in the cliffs near the Elsa Gate (Plate 16.4a). An iconic feature of Hell's Gate is the occurrence of rock pinnacles, including the Fisher's and Central Towers. These are erosional relics of retreating cliff faces and are comprised of columnar jointed ignimbrite, not volcanic plugs as has been reported. The park also includes the opportunity to examine caves and small overhangs, sections of collapsed lava tubes.

16.5.2   Njorowa Gorge

The Njorowa Gorge is a narrow slot up to 200 m deep, with near-vertical walls (Plate 16.1). The gorge contains an ephemeral stream that flows southwards, away from Lake Naivasha (Fig. 16.1b). The gorge was formed by waters which flowed out of the palaeo-Lake Naivasha during the high stands of the Late Pleistocene and Early Holocene. The bulk of the eruptions at Olkaria probably occurred beneath this palaeo-lake. The river is not currently connected to the lake, as may be thought, but is fed mostly by internal springs. The walls of the gorge reveal superb exposures of finely bedded rhyolitic ash, indicative of the subaqueous eruptions (Plate 16.4b). Flow banding and pipe vesicles can be observed in the rhyolitic lavas. The gorge also includes several fumaroles and hot springs which are associated with multicoloured algae.

16.5.3   Obsidian

Obsidian (also known as volcanic glass) is a ubiquitous component of rhyolitic volcanism and is abundant at both Olkaria and Yellowstone. The jet black appearance and glassy texture, with pronounced conchoidal fracture pattern, is characteristic (Plate 16.5a). The obsidian at Olkaria occurs in several localities, including an unusually large vein in the Njorowa Gorge. It is also found in overhangs or caves associated with lava tubes. The fist-sized lumps found on the floor of the gorge have eroded from veins (Plate 16.5b). The rhyolite in the sidewalls of veins may contain thin bands of obsidian. Obsidian was commonly used by ancient cultures for stone tools, e.g. at Yellowstone and in Ethiopia, as small pieces readily shatter into knife-sharp shards. The archaeological sites at Olorgesailie, between Nariobi and Lake Magadi, probably contain the greatest number of stone tools ever discovered. The source of the obsidian has not been established, but Olkaria is a possibility.

Obsidian forms when the mobility of ions within a highly siliceous or rhyolitic melt is prevented from achieving an ordered crystalline pattern. This usually results from very rapid cooling (Bakken 1977). Obsidian is associated with flows that have a lower viscosity than their rhyolite host, thus the banded texture. Typically, the obsidian forms as one of the final eruptive phases, after the bulk of the gas and pumice have been vented.

16.6 Hiking Trails

The main tourist attraction in the Mount Longonot National Park is a 13.5-km-long hiking trail that provides a relatively easy opportunity to climb an active volcano. The trail starts from the main gate at an altitude of 2,150 m. The highest point on the cone is a small peak on the northwestern rim of the crater. The rim of the crater includes some scrambles on recent deposits of ash and pyroclastics. Small fumaroles and sulphur vents can be observed from the main track. The view southwards includes evidence of an older caldera, as well as of the Suswa Volcano. In 2009, an intense bushfire caused considerable damage and some photographs of Longonot depict a barren scene, whereas, typically, the slopes of the central cone are well wooded. The hike along the upper parts of the Njorowa Gorge in the Hell's Gate National Park, rimmed in by near-vertical cliffs, provides an opportunity to examine a section of a rhyolitic volcano. It is intriguing to compare differences with the basaltic volcanism associated with the Longonot Volcano. Erosion has littered the floor of the gorge

with samples of obsidian that range in size from boulders and pebbles to gravel. The Njorowa Gorge is subject to (rare) flash floods, and in recent years several escape routes have been organised. The gorge has been the setting of a number of internationally known action films. The park has a Joy Adamson educational centre, which together with the Elsa Gate is named in memory of this famous conservationist.

16.7 Geothermal Fields

The potential for geothermal energy in the rift valley of central/southern Kenya has long been known, and many of the active, Holocene volcanoes (Sect. 6.7) have substantial resources. The geothermal energy of volcanic terranes surrounding Lake Naivasha has been discussed by Clarke et al. (1990), with details of the Mount Longonot Volcano by Alexander and Ussher (2011). The resource at Olkaria was confirmed by drilling in 1967, although the first investigations date from 1955. Seismic and gravity methods have been used to evaluate the depth and extent of magma chambers in the Naivasha basin, as well as in the adjacent Nakuru basin (Simiyu 1996).

The Olkaria geothermal field has been described by Simiyu (1996) and Simiyu and Keller (2000) as a world-class power generation site with a high enthalpy field that covers an area of 240 km^2. The current operator, KENGEN, drilled the first set of wells in 1976. Wells are drilled to a depth of 750–1,300 m where they source temperatures of up to 340 °C. A fourth steam turbine-driven generating station has recently been commissioned and the field currently produces some 600 MW of geothermal power. The resource has an estimated potential of 2,000 MW, almost double the peak demand for Kenya. The hydrothermal waters are relatively siliceous and, together with other impurities, may block wells and turbines. The view of a series of wellheads venting off steam under pressures too great to be piped may not appeal to all environmentalists (Plate 3.3), but Olkaria is an important resource of relatively clean energy.

Plate 16.2   a The main cone of Mount Longonot viewed from the northern slopes; **b** The thickly forested summit crater of the Longonot Volcano appears as a lost wilderness

Plate 16.3 a The exposed rim of the caldera viewed from the southern side of the main cone, Mount Longonot. Deeply eroded, pale volcanic ash is visible in the foreground; **b** Pale coloured pyroclastics with abundant lithic fragments can be observed on the rim of the summit crater, Mount Longonot

Plate 16.4 **a** Columnar jointed ignimbrite at the base of the cliffs near the Elsa Gate, Hell's Gate; **b** Fine-scale layering of rhyolitic volcanic ash indicative of a subaqueous origin, Njorowa Gorge, Hell's Gate

Plate 16.5 Obsidian is abundant in Njorowa Gorge, occurring within a finely banded rhyolite (**a**) and in fist-sized lumps eroded from a large vein (**b**)

References

Alexander, K. and Ussher, G. (2011). *Geothermal resource assessment for Mt Longonot, Central Rift Valley, Kenya.* Nariobi: Proceedings of the Kenya Geothermal Conference.

Baker, B. H. (1987). Outline of the petrology of the Kenyan Rift alkaline province. In: Fitton, J. G. & Upton, B. G. J. (eds.), *Alkaline igneous rocks* (Vol. 30, pp. 293–311). Geological Society of London Special Publication.

Bakken, B. (1977). Obsidian and its formation. *North West Geology, 6–2,* 88–92.

Clarke, M. C. G., Woodhall, D. G., Allen, A., & Darling, W.G. (1990). *Geological, volcanological and hydro-geological controls on the occurrence of geothermal activity in the area surrounding Lake Naivasha, Kenya.* Report of Ministry of Energy, Kenya and British Geological Survey.

Scott, S. C. (1980). The geology of Longonot volcano, Central Kenya: a question of volumes. *Philosophical Transactions of the Royal Society of London, A296,* 437–465.

Scott, S. C., & Skilling, I. P. (1999). The role of tephra chronology in recognizing synchronous caldera-forming events at Quaternary volcanoes Longonot and Suswa, south Kenya Rift. In: Firth, C. R. & McGuire, W. J. (eds.), *Volcanoes in the Quaternary* (Vol. 161, pp. 47–67). Geological Society of London Special Publication.

Simiyu, S. M. (1996). An integrated application of seismic and gravity methods to magma chambers beneath the Nakuru-Naivasha sub-Basin. *EOS, Transactions American Geophysics Union, 76,* 257–258.

Simiyu, S. M., & Keller, G. R. (2000). Seismic monitoring within the Olkaria geothermal field. *Journal of Volcanology and Geothermal Research, 95,* 197–208.

Woolley, A. (2001). *Alkaline rocks and carbonatites of the world. Part 3: Africa* (372 p). Geological Society of London.

Lake Natron and the Oldoinyo Lengai Volcano

17

Abstract

Lake Natron is located in a desolate, relatively low-lying section of the Gregory Rift Valley, northern Tanzania. The lake is bordered to the west by the Western Escarpment, a major rampart constructed of layer upon layer of Pliocene-age volcanic lavas and tephra, and to the east by the giant Gelai Volcano. Natron is an extraordinarily toxic, soda lake (pH of 12) with extensive sodium carbonate salt flats. The symmetrical cone of Oldoinyo Lengai (3,188 m) rises some 2.5 km above the valley floor to the southwest of Lake Natron. Oldoinyo Lengai is one of the most famous Holocene volcanoes on Earth as it was here the coexistence of two immiscible magmas with radically different compositions was first observed in nature. The alkaline silicate magmas are the most abundant, but it is the highly unusual natrocarbonatite magmas for which Lengai is so well known. The natrocarbonatite is erupted at temperatures considerably lower than silicate lavas. The high sodium contant of the natrocarbonatitite lavas and ashes reacts rapidly with meteoric waters to create secondary, white coloured minerals within a few days. The altered, white or pale grey natrocarbonatite ashes that partially cloak the upper slopes resemble an ice cap when seen from distance. Flows of natrocarbonatite associated with eruptions from the early 1990s could formerly be observed in a shallow summit crater which had remained intact since 1960. This feature overflowed and collapsed during violent eruptions between 2007–2008 and the new crater is too deep and unstable to enter. Lengai erupts on average every 15–20 yr, although some of the recent eruptions have been of nephelinite rather than natrocarbonatite. Lengai presents a potential hazard to the local Maasai who live in the proximity of Lake Natron, and also to aircraft as ash columns may attain heights of tens of kilometres.

Keywords

Active volcano • Natrocarbonatite • Nephelinite
Sodic brines • Strombolian eruptions • Summit crater

Photographs not otherwise referenced are by the author.

© Springer International Publishing AG, part of Springer Nature 2018
R. N. Scoon, *Geology of National Parks of Central/Southern Kenya and Northern Tanzania*, https://doi.org/10.1007/978-3-319-73785-0_17

Plate 17.1   The spectacular cone of Oldoinyo Lengai rises abruptly from the floor of the Gregory Rift near Lake Natron

17.1   Introduction

The Gregory Rift Valley in the far northern part of Tanzania is a desolate, arid location that is protected in a number of conservancies run in partnership with the local Masaai people. The minimum elevation is 600 m and vegetation is sparse, rainfall erratic, and temperatures are regularly over 40 °C. The dominant features are Lake Natron, one of the largest of the ribbon-shaped, alkaline lakes in the rift valley giant volcanic cones (Fig. 17.1). The most spectacular of these is the nearly perfectly symmetrical cone of the Oldoinyo Lengai Volcano (Plate 17.1). The symmetry is consistent with a dominantly Holocene-age volcano that has undergone minimal erosion. In comparison, the Pliocene- and Pleistocene volcanoes of the region, which include Shombole (north of the lake), Gelai (east of the lake), Mosonik (perched on top of the escarpment) and

Kerimasi (south of Lengai), are severely eroded. Lake Natron is named after the extensive salt deposits that occur during all but the wettest periods. Natron is a type of sodium carbonate which when mixed with trona is the principal component of deposits formed by the sodic brines (Sect. 15.7). The lake is bordered to the west by the Western Escarpment, a major rampart constructed of layer upon layer of Pliocene-age volcanic lavas and tephra (Frontispiece 1). Tectonism associated with movement on the fault associated with the Western Escarpment, which occurred during the Pleistocene at approximately 1.2 Ma, triggered the eruption of many of the volcanoes in this region (Dawson 2008).

The name 'Oldoinyo Lengai' translates to 'Mountain of God' from the Maasai language and is thought to have arisen because the pale grey-white lavas and ashes that cloak the upper slopes resemble an icecap when seen from a distance.

These lavas and ashes are products of the unusual natro-carbonatite volcanism, first identified and described by Dawson (1962a). The lower slopes are dominated by nephelinite and phonolite ashes. Oldoinyo Lengai is the most active of the Holocene volcanoes in the Gregory Rift (Dawson 1962b; Dawson and Mitchell 2008).

17.2 Volcanism of the Region

A number of large (extinct) volcanoes occur both within this section of the Gregory Rift Valley and in the Eyasi Half-graben (Figs. 5.1 and 5.5), as reviewed by Woolley (2001). The eroded cone of the Mosonik Volcano is perched on the crest of the Western Escarpment above Lake Natron. This volcano is rather poorly known, although Dawson (2008) presented some new field evidence. There are two conflicting radiometric age dates, both based on nephelinite lava flows, a Pliocene age of 3.12 Ma (Isaac and Curtis 1974) and a Pleistocene age of 1.28 Ma (Forster et al. 1997). The Shombole Volcano, which is located in the rift valley to the north of Lake Natron, Kenya, has a Pliocene age of 2.0–1.96 Ma (Fairhead et al. 1972). This overlaps with the Ngorongoro Volcanic complex.

The rugged, deeply eroded cone of the Gelai Volcano, located to the east of Lake Natron, was formerly thought to be an extinct Pleistocene feature. The age of 0.99–0.96 Ma has been determined from samples of nephelinite (Evans et al. 1971). In 2007, however, small fissures opened on the lower slopes. They have been related to a volcano-tectonic event at depth, probably intrusion of a small dyke (Delvaux et al. 2008).

The Kerimasi Volcano is located to the south of Oldoinyo Lengai and is similar in that it includes both nephelinite and natrocarbonatite lavas and ashes (Dawson 2008; Mattson and Kervyn 2014). The main phase of volcanism has a Pleistocene age of approximately 0.6–0.4 Ma, associated with the final stage of rift faulting in this area (Hay 1976), although there has been younger activity which has not been dated. Ashes from eruptions of Kerimasi younger than 0.4 Ma are found at Oldupai Gorge (Sect. 11.5). The cone of the Kerimasi Volcano was severely eroded during the wetter and humid climates of the Pleistocene and Early Holocene.

The explosion craters and tuff cones in the rift valley to the southeast of Kerimasi (not shown in Fig. 17.1) are collectively known as the Engaruka–Natron field (Dawson and Powell 1969). They encompass a broad range of ages from 0.57 Ma to as young as 0.14 Ma (MacIntyre et al. 1974). The locality known as God's Pit (Shimo la Mungu in the Maasai language), which is visited by some tourists, is an interesting example of a small explosion crater in which the volcanic strata in the sidewalls can be observed to be downwarped.

17.3 Lake Natron

Lake Natron has a maximum length of 57 km and width of 22 km, although the size varies considerably. The lake is extremely shallow, with a maximum depth of 3 m, and is exceptionally alkaline (average pH of 12). During the Early Holocene, Lakes Natron and Magadi were connected to form a much larger palaeo-lake (Fig. 6.1). The composition of the sodic brines and salt deposits of Natron are similar to those associated with Lake Magadi. In both lakes, animal and bird

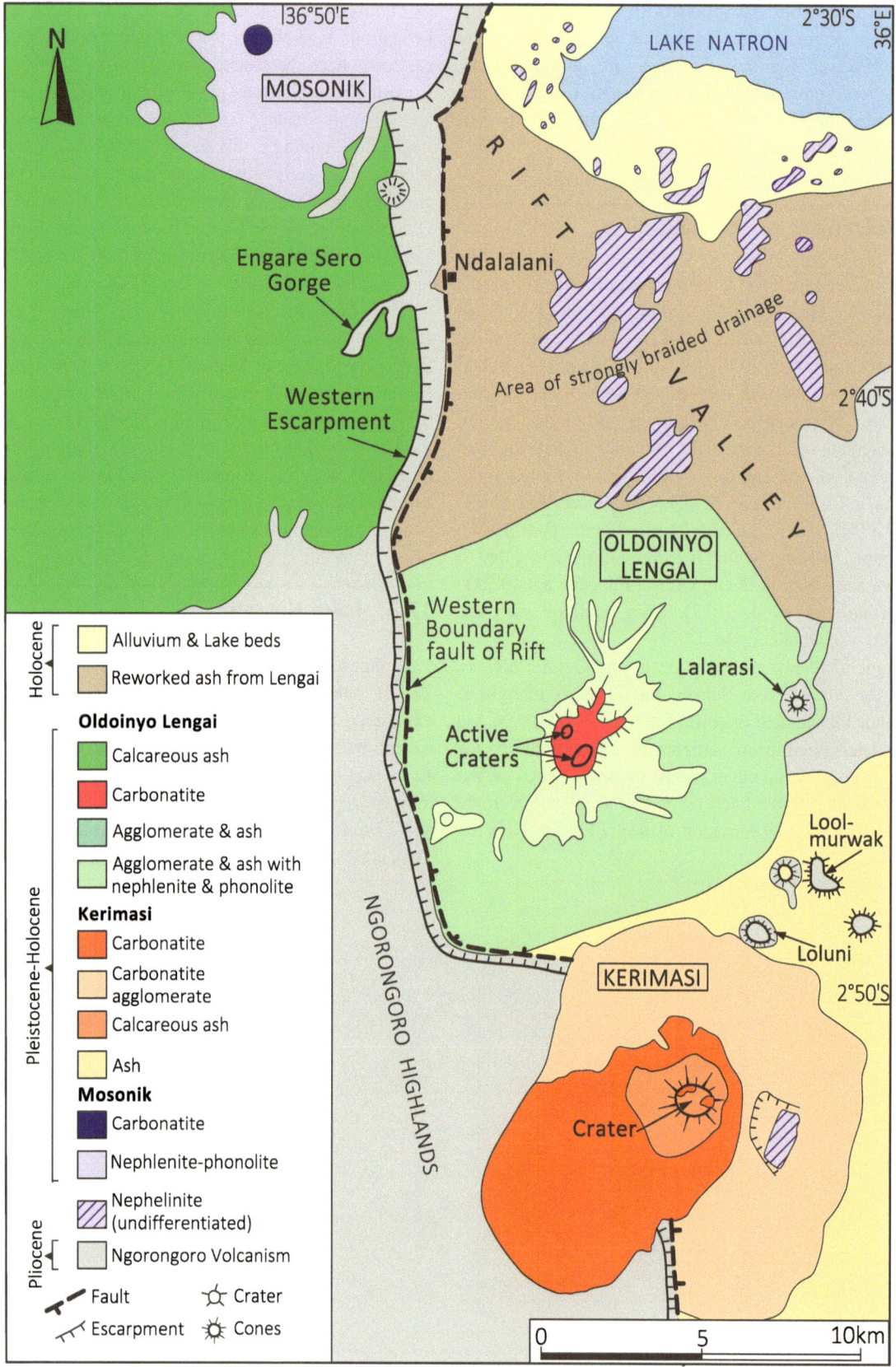

Fig. 17.1 Simplified geological map of the area around Oldoinyo Lengai based on the Quarter Degree Sheet 39, Angata Salei (1961)

remains are preserved and fossilised by sodium carbonate salts. The extreme toxicity results in high concentrations of cyanobacteria that are sometimes visible in satellite images (Fig. 17.2). The Natron basin is relatively large and includes the Southern Ewaso Ng'iro River that rises in southern Kenya. The extreme aridity of the region, however, results in evaporation far exceeding inflow. The sodic nature of the brines is further enhanced by weathering of the sodium-rich volcanoes, both of nephelinite and natrocarbonatite compositions.

Lake Natron contains extensive salt deposits which have been prospected for production of soda ash. The Kenyan section of the Natron basin has been targeted for a hydroelectric scheme, although the biodiversity has led the WWF to recognise the region as a unique halophytic ecosystem. Near the southwestern shore, Lake Natron is fed by small freshwater streams, with ponds and springs (Plate 17.2a). These features sustain a wide range of aquatic birds as well as large game, including zebra and giraffe. An interesting observation is that the freshwater and alkaline sections of the lake rarely mix. A walk onto the salt flats in this area reveals the extensiveness of the sodium carbonate salt deposits as well as the characteristic hexagonal cracks (Fig. 17.2b). The extreme alkalinity of large parts of the lake, as well as the remote setting, has resulted in Natron being the dominant breeding ground for flamingo in East Africa (Sect. 15.10).

17.4 Oldoinyo Lengai

Oldoinyo Lengai is the most well known example of the remarkable bimodal alkaline basalt and natrocarbonatite volcanism that characterises parts of the Gregory Rift Valley (Dawson 2008). The extinct Kerimasi Volcano has a similar petrogenesis, whereas most of the other Pleistocene volcanoes within this area are grouped with the alkaline basaltic trend. Oldoinyo Lengai poses a potential hazard to aircraft as the mountain is located close to the flight path between Arusha and Nairobi, as well as transcontinental flight paths between Southern Africa and Europe. An observation made by some visitors is the ejection of near-vertical columns of intensely hot, brownish gases that jet into the stratosphere. J. B. Dawson observed major eruptions in 1966–1967, including the one on 14 August 1966, in which ash clouds attained a height of approximately 3,000 m.

17.4.1 Historical Activity

Oldoinyo Lengai has been active since the Late Pleistocene (oldest activity is estimated at 0.37 Ma). Significant eruptions are estimated to have occurred during the Holocene. The history of the eruptions has been documented on websites, e.g. St. Lawrence University, Canada. The first eruption

Fig. 17.2 The unusual colour of the northern part of Lake Natron in this satellite image is due to the red photosynthesising pigment in the cyanobacteria. *Source* NASA Terra-ASTER image for 2003, courtesy of Philip Eales, Planetary Visions

of Lengai to be witnessed and reported was in 1883. A major eruption in 1917 killed off most of the trees on the outer slopes, and deposited a thick blanket of grey ash over large areas. This ash was partially covered in some areas by a much darker, poorly consolidated ash from major eruptions during 1940–1941. This latter event may have formed the dunes in the Ngorongoro Conservation Area (NCA) known as the 'Shifting Sands' (Sect. 10.8). The major eruptions of 1966–1967 excavated two summit craters, the largest of which had a diameter of 400 m and depth of 150 m.

The larger eruptions of Lengai have resulted in thick deposits of ash being dispersed over the NCA due to the prevailing easterly winds (Dawson 1964), e.g. the Holocene ash which covers parts of the Ngorongoro Caldera (Fig. 10.3). Ashfall has also spread onto the Eastern Serengeti Plains (Fig. 7.1). The ash fall from the 1966–1967 eruptions reached as far as the Serena Lodge, some 150 km from Lengai. Moreover, Bed VII of the Oldupai Group is comprised of ash related to an eruption of Lengai with an age of 22,000–15,000 BP (Sect. 11.5).

17.4.2 Petrology

The most fascinating component of the Oldoinyo Lengai volcanism is that the two magma types, the alkaline silicate and natrocarbonatite coexist as immiscible melts (Dawson et al. (1995a). The average composition of the main cone is not known as the core is not exposed (due to its youth), but it is widely accepted that the alkaline silicate magmas, which erupt as phonolite and nephelinite, are dominant (Dawson 2008). The natrocarbonatite magmas are distinctly subordinate and they are thought to have been absent during the early history of the volcano. The oldest natrocarbonatite lavas and ashes have an age of approximately 1,250 BP. The main cone is dominated by yellow-coloured phonolitic and nephelinitic tuffs and agglomerates, indicative of the explosive style of the alkaline silicate volcanism. Dark-coloured nephelinite and phonolite lava flows are considerably less common than the tephra. They can be observed on the outer flanks. A number of pyroclastic units can also be identified. Plutonic blocks of jacupirangite-ijolite and nepheline syenite are also recognised (Dawson et al. 1995b). The natrocarbonatite is observed as the distinctive, pale grey deposits on the upper flanks of the cone (Plate 17.3a). The natrocarbonatite ashes can also be observed at the base of the mountain near Lake Natron.

17.4.3 Natrocarbonatite

The unusual, sodium-rich variety of carbonatite that characterises both Oldoinyo Lengai and Kerimasi has only been identified in volcanoes of the Gregory Rift (Dawson 2008; Woolley 2001). Extinct carbonatite volcanoes and intrusive complexes from other localities are associated with the calcic variety of carbonatite. The high content of sodium carbonate in the natrocarbonatite is balanced by a relatively paucity of potassium and calcium. An interesting aside is that even though the extinct calcic variety of carbonatite volcanism is dominated by calcium carbonate, the possibility that they may originally have been comprised of natrocarbonatite (that oxidised on contact with the atmosphere, with the sodium being replaced by calcium, as deduced by evidence from Oldoinyo Lengai), cannot be dismissed. A characteristic feature of the natrocarbonatite melts, as witnessed at Oldoinyo Lengai, is their extremely low viscosity (ten times less than basalt and possibly the lowest reported for terrestrial lavas) and very low eruption temperatures (approximately 500 °C). This results in fast-travelling lavas that freeze as anomalously thin flows.

The high sodium content of the lavas and ashes reacts rapidly with meteoric water to form secondary phases, thus despite being erupted as black flows, the natrocarbonatite oxidises to light-coloured secondary minerals within a few days. The primary mineral within the natrocarbonatite is *nyerereite*, a sodium-calcium carbonate identified by

Dawson (1962b) and named after Tanzania's first president. This mineral is only preserved in the freshly erupted lavas. The sodium in the nyerereite is rapidly replaced by calcium to form calcium carbonate. Other new minerals discovered in the natrocarbonatite volcanics of the Gregory Rift are *gregoryite*, a potassium carbonate, and *kerimasite*, a rare garnet. The natrocarbonatite lavas and ashes contain a relatively high proportion of radioactive minerals and rare earth elements (REE).

17.4.4 The Summit Crater (1960–2006)

The summit crater first described in the 1960s included northern and southern components, of which the latter was smaller and rather difficult to access (Dawson 1962a, b). In November 1988, small-scale eruptions of natrocarbonatite lava were observed that thermally eroded the floor of the northern crater (Dawson 1995a). By 1992, the wall of this crater had been reduced to a height of only 15 m (Fig. 6.2b). The flat floor was attributed to repeated eruptions of small quantities of lava. Hornitos and spatter cones were reported as regularly ejecting ash and lava blocks. Small eruptions and persistent Strombolian-style events in 1993 included flows of natrocarbonatite lava.

During a visit by the author in May 1995, natrocarbonatite ashes were observed to envelop large parts of the steep, upper slopes of the main cone (Plate 17.3b) and the rim of the northern crater (Plate 17.4a). Small ash and spatter cones were a significant feature of the flat-floored crater. Active vents could be identified by the sound of magma in the conduits (Plate 17.4b). The flows of natrocarbonatite from the 1993 eruption were observed as shiny white, smooth-topped, *pahoehoe* (ropy-textured) flows which resembled tongues of travertine rather than lava (Plate 17.5a). The pahoehoe flows were overlain by *aa* (blocky textured) flows that had been reduced to crumbly, pale grey debris. The thinness of the 1993 flows (approximately 5–10 cm) is a function of the low viscosity. Channels could be identified in the 1993 flows by physical boundaries with a higher heat flow indicative of migration of interstitial melt beneath the carapace (Plate 17.5b). Fumaroles in the summit crater were observed as being associated with prominent mounds of geyserite. They also contained crystals of yellow sulphur in small vugs. One of the hazards of the crater, in addition to toxic gases, was subsidence and unstable overhangs near some vents.

During December 1995, a team from Geneva filmed Strombolian-style eruptions within the northern summit crater. In 1998, spatter cones were reported as still being active and some lava overflowed the rim of the crater. In

1999, natrocarbonatite lava was erupted from a small vent on the outer (eastern) slopes, and lava was reported as having overflowed the crater on the western side. This sporadic activity continued during 2001–2002. By 2006, the northern crater was reported as having filled up, with new lava flows regularly spilling over the rim. One of the 2006 flows extended down the outer flanks for a lateral distance of 3 km.

17.4.5 Recent Activity (2007–2016)

Major seismic tremors, some of which were felt as far afield as Nairobi, occurred between 12 and 18 July 2007. This culminated in a major eruption on 4 September 2007 that continued intermittently over several weeks. Lava regularly overflowed the summit crater and a major flow was observed low down on the northwestern flanks. Approximately 1,000,000 m^3 of natrocarbonatite is estimated to have been erupted during this event. The emplacement and inflation of natrocarbonatite lava flows during these eruptions was described by Mattson and Vuorinen (2009). They identified lava with both *aa* and *pahoehoe* textures. The different physical forms were found to be dependent on effusion rates, rather than compositional differences. Additional activity during 2007, as well as throughout 2008, marked a change from the quiescent effusive activity that had characterised the eruption of natrocarbonatite flows since 1960. This was replaced by explosive activity that marked a change to eruption of the undersaturated alkaline silicate magmas (Mattson and Vuorinen 2009). In 2010, two lava flows were erupted and a small lava lake was reported to be present in the crater. In 2013, a sequence of particularly violent eruptions occurred. The 2007–2013 eruptions resulted in the 1960 summit craters collapsing downwards and the new crater is several hundred metres deep. The walls are too steep and unstable to enter (Plate 17.6a). Oldoinyo Lengai has been quiescent since 2014 and during a visit in September 2016, no activity was apparent.

17.5 Deposits in the Rift Valley

The floor of the Gregory Rift Valley to the southwest of Lake Natron is mantled by a thin dusting of black (phonolite-nephelinite) volcanic ash, mostly derived from Oldoinyo Lengai. Very pale grey deposits of ash related to Holocene eruptions of natrocarbonatite also occur. A detailed remote sensing study of this area has led to reassessment of the volume of material thought to have erupted from Oldoinyo Lengai, with a new, much larger estimate of 41 ± 5 km^3 (Kervyn et al. 2008). This is due to the recognition of numerous debris avalanche deposits (DADs) located in this region of the valley (Box 13.1). They are ascribed to Oldoinyo Lengai and Kerimasi. DADs are observed as hummocky, irregular ground that includes low, linear ridges, such as those traversed by the main access road. The abundance of DADs may suggest that the cone of Oldoinyo Lengai is a relatively recent feature. Older cones may have collapsed during the Plinian-style eruptions that are required to generate these features.

17.6 Trekking and Footprints

The trek to the summit of Oldoinyo Lengai is generally completed in one day as there are no huts or established camp sites. Even with a predawn start, the ascent is strenuous and a guide is recommended as the flanks of the cone reveal radial erosion features with steeply incised, U-shaped gullies. The gain in altitude is considerable when the low elevation of the rift valley is taken into account. Much of the ascent is undertaken on deeply weathered scree and volcanic ashes with only a few sections on rock outcrops. Note that access into the summit crater is no longer possible. Fossilised footprints of *Homo sapiens* have recently been located in the area south of Lake Natron (Plate 17.6b). They can be reached by a short hike over sequences of volcanic ash derived from Oldoinyo Lengai. The footprints have been dated at 120,000 yr but this needs reassessing (10,000 yr maximum?) on the basis of local knowledge of the Maasai guides (who note cattle prints and historical movement of their people). Moreover, the footprints are preserved in an ash layer close to the surface that is probably Holocene. Additional highlights include a walk onto the salt flats that surround Lake Natron, as noted above, as well as the scenic waterfalls of the Engare Sero Gorge which includes a scramble on the rock walls of the gorge and wading in the cool river. The walls of the gorge reveal excellent exposures of the Ngorongoro Volcanic complex, including lava flows, ash layers, debris deposits and breccia dykes (Plates 5.2–5.4).

Plate 17.2 a Natron includes freshwater streams and ponds on the southwestern fringe of an otherwise highly toxic lake. The two water types do not mix readily. The view includes the Gelai Volcano in the far distance; **b** Mud cracks in sodium carbonate salt deposits reveal characteristic hexagonal patterns, Lake Natron

Plate 17.3 **a** Pale grey natrocarbonatite ash mantled the upper flanks of Oldoinyo Lengai beneath the active northern crater in 1995; **b** Ascent of Oldoinyo Lengai in 1995 included scrambling on partially consolidated volcanic ash

Plate 17.4 **a** The flanks of the northern crater of Oldoinyo Lengai consisted of layers of pale grey and white natrocarbonatite ash in 1995; **b** The northern crater of Oldoinyo Lengai included active vents in 1995

Plate 17.5 a Thin pahoehoe (ropy-textured) flows of natrocarbonatite lava from the 1993 eruption in the northern crater of Oldoinyo Lengai were observed in 1995. The overlying aa (blocky) flow was severely weathered and broken (due to volume change associated with formation of the secondary minerals) in comparison to the smooth-looking pahoehoe flows; **b** A channel in one of the 1993 pahoehoe flows of natrocarbonatite lava in the northern crater of Oldoinyo Lengai, was notably hotter in comparison to the outer parts of the flows when observed in 1995

Plate 17.6 a The summit crater of Oldoinyo Lengai is currently inaccessible. *Source* Public domain website http://www.natgeocreative. com/comp/DD/001/1399852.jpg; **b** Fossilised footprints of *Homo* *sapiens* located to the south of Lake Natron occur in volcanic ash (natrocarbonatite) erupted from the Oldoinyo Lengai Volcano

References

Dawson, J. B. (1962a). Sodium carbonatite lavas from Oldoinyo Lengai, northern Tanganyika. *Nature, 196,* 1065–1066.

Dawson, J. B. (1962b). The geology of Oldoinyo Lengai. *Bulletin of Volcanologique, 24,* 349–387.

Dawson, J. B. (1964). Carbonatitic volcanic ashes in northern Tanganyika. *Bulletin of Volcanologique, 27,* 81–92.

Dawson, J. B. (2008). *The Gregory Rift Valley and Neogene-recent volcanoes of northern Tanzania* (Vol. 33, 102 p). Geological Society London Memoir.

Dawson, J. B., & Mitchell, R. H. (2008). Oldoinyo Lengai. *Contribution to Bulletin of the Global Volcanism Network, 32,* 14–15.

Dawson, J. B., & Powell, D. G. (1969). The Engaruka-Natron explosion crater area, northern Tanzania. *Bulletin of Volcanologique, 33,* 791–8172.

Dawson, J. B., Pinkerton, H., Norton, G. E., Pyle, D. M., Browning, P., Jackson, D., et al. (1995a). Petrology and geochemistry of Oldoinyo Lengai lavas extruded in November 1988: Magma source, ascent, and crystallization. In K. Bell & J. Keller (Eds.), *Carbonatite volcanism* (pp. 251–283). Berlin: Springer.

Dawson, J. B., Smith, J. V., & Steele, I. M. (1995b). Petrology and mineral chemistry of plutonic igneous xenoliths from the carbonatite volcano, Oldoinyo Lengai, Tanzania. *Journal of Petrology, 36,* 797–826.

Delvaux, D., Smets, B., Wauthier, C., Macheyeki, A. S., D'Oreye, N., Oyen, A., et al. (2008). Surface ruptures associated to the July–August 2007 Gelai volcano-tectonic event, North Tanzania. In *Geophysical Research Abstracts* (10 EGU2008-A-06954).

Evans, A. L., Fairhead, J. D., & Mitchell, J. G. (1971). Potassium-argon ages from the volcanic province of northern Tanzania. *Nature Physical Science, 229,* 10–20.

Fairhead, J. D., Mitchell, J. G., & Williams, L. A. J. (1972). New K/Ar determinations on rift valley volcanoes of South Kenya and their bearings on age of rift faulting. *Nature Physical Science, 238,* 66–69.

Foster, A., Ebinger, C., Mbede, E., & Rex, D. (1997). Tectonic development of the Northern Tanzania sector of the East African Rift System. *Journal of the Geological Society of London, 154,* 689–700.

Hay, R. L. (1976). *Geology of the Olduvai Gorge: A study of sedimentation in a semiarid basin* (p. 203). Berkeley: University of California Press.

Isaac, G. L., & Curtis, G. H. (1974). Age of the Acheulian industries from the Peninj Group, Tanzania. *Nature, 249,* 624–627.

Kervyn, M., Ernst, G. G. J., Kladius, K., Keller, J., Mbede, E., & Jacobs, P. (2008). Remote sensing study of sector collapses and debris avalanche deposits at Oldoinyo Lengai and Kerimasi volcanoes, Tanzania. *International Journal of Remote Sensing, 19,* 6565–6595.

MacIntyre, R. M., Mitchell, J. G., & Dawson, J. C. (1974). Age of fault movements in Tanzanian sector of East African rift system. *Nature, 247,* 354–356.

Mattson, H. B., & Kervyn, M. (2014). *Insights into a carbonatite volcano: Kerimasi, Northern Tanzania* (Abstract A-001). Edinburgh: Volcanic and magmatic studies group.

Mattson, H. B., & Vuorinen, J. (2009). Emplacement and inflation of natrocarbonatite lava flows during the March–April 2006 eruption of Oldoinyo Lengai, Tanzania. *Bulletin of volcanology, 71,* 301–311.

Quarter Degree Sheet 39 Angata Salei (1961). *Including a brief explanation of the geology with mapping by Guest, N. J. (1950), James, T. C. (1952), Pickering, R. (1957) and Dawson, J. B. (1960).* Geological Survey of Tanganyika.

St. Lawrence University, Canada. http://blogs.stlawu.edu/lengai/.

Woolley, A. (2001). *Alkaline rocks and carbonatites of the world. Part 3: Africa* (372 p). Geological Society of London.

About the Author

Dr. Roger N. Scoonwas born in Bristol, England, and grew up in a rural village in North Somerset. After obtaining bachelors (1976) and masters (1976) degrees in the Department of Mineral Exploitation at University College, Cardiff, he worked for 3 years as a mine geologist in Zambia. A doctorate at Rhodes University, Grahamstown (1985) initiated a lifetime passion for the platinum and chromite deposits of the Bushveld Complex, South Africa. Apart from a short stint as a field geologist in Namibia, Roger spent the bulk of his career as an exploration geologist in the Bushveld Complex, including as a principal and director of Platexco Incorporated, a Toronto-listed junior explorer. The successful Winnaarshoek project (Marula mine), where Roger was responsible for surface mapping, refurbishing and sampling old mine workings, running an extensive drilling program, and the bankable feasibility was purchased by Impala Platinum in 2001. Roger has published a number of scientific articles in local and international journals, mostly on the Bushveld Complex. Visits to the national parks of East Africa over many years, including trekking on the giant volcanic cones led to an interest in the East Africa Rift System. Short descriptions of geosites from localities around the world, including East Africa, published in the quarterly journal, The Geobulletin, since 2010 have led to a new interest as 'The Geotraveller'. Roger has led many geological field trips, most recent were to the Eastern Limb of the Bushveld Complex and National Parks of northern Tanzania for the 35th International Geological Congress (2016). Roger is a Senior Researcher at Rhodes University and has acted as an external examiner for the Masters programme in Exploration Geology since 1988. He is a Fellow of the Geological Society of South Africa (FGSSA), the Society of Economic Geologist (FSEG) and the Geological Society of London (FGS). Other passions include skippering the family yacht, 'Makoma' in the Eastern Mediterranean, as well as football, bridge and wildlife photography. Roger is a South African citizen and lives with his wife Amelia in the largely unspoilt Garden Route of the Western Cape.

© Springer International Publishing AG, part of Springer Nature 2018
R. N. Scoon, *Geology of National Parks of Central/Southern Kenya and Northern Tanzania*, https://doi.org/10.1007/978-3-319-73785-0

Glossary

Agglomerate Coarse-grained volcanic breccia, i.e. a rock comprised of angular fragments of pyroclastic material, typically found near a vent.

Albertine Rift The Western Branch of the EARS, named after Lake Albert in the northern part of the structure.

Alkali or Alkaline Any strongly basic substance, such as a hydroxide or carbonate of one of the alkali metals such as sodium (Na) or potassium (K).

Alkali Basalt A group of volcanic rocks strongly undersaturated in silica and which contain more alkali metals (particularly sodium and potassium) than considered average for basalts.

Alkaline Lake See Soda Lake.

Alkaline Rocks Volcanic or intrusive igneous rocks which contain sufficient sodium and potassium for alkaline minerals (e.g., alkali pyroxene, alkali amphibole, feldspathoids such as nepheline) to crystallise. Alkali rocks make up less than 1% by volume of exposed igneous rocks.

Alkaline Magma Magma unusually rich in Na_2O and K_2O which is discriminated from the far more common subalkali magmas.

Anticline An upfolded structure in which the oldest rocks occur in the core and the youngest rocks in the outer part.

Basalt A general term for dark-coloured (melanocratic), mafic igneous rocks dominated by extrusive lavas and ashes. The dominant minerals are calcic plagioclase and clinopyroxene. Basalts are subdivided using chemical discriminations such as the TAS diagram.

Basanite Part of the basaltic group of igneous rocks characterised by being silica-undersaturated and containing a feldspathoid such as nepheline.

Basement A general term to describe the rock sequences, possibly undifferentiated and typically older than the near-surface rocks that underlie the region of interest. In many areas, including East Africa, the basement is dominated by crystalline igneous and metamorphic terranes.

Basement Terrane The lowest (and typically the oldest) of the mappable rock units in the region of interest.

Basin A low-lying, or depressed area, often downfolded, with no surface outlet, such as a lake basin or a groundwater basin.

Block Faulting The dominant structure resulting from an area subjected to extensional tectonism, i.e. including parallel faults which may develop to include a series of grabens and horsts.

Brine Highly saline fluids, typically warm or hot, rich in alkali and related elements such as calcium, sodium, potassium and chlorine that are generally only very minor constituents of water.

Calcareous Pertaining to material dominated by calcium carbonate.

Calcrete Surficial deposits of sand and gravel cemented by calcium carbonate precipitated from infiltrating groundwater.

© Springer International Publishing AG, part of Springer Nature 2018
R. N. Scoon, *Geology of National Parks of Central/Southern Kenya and Northern Tanzania*, https://doi.org/10.1007/978-3-319-73785-0

Caldera A large, basin-shaped volcanic depression, more or less circular, formed by a catastrophic process generally associated with collapse of the *entire* cone induced by rapid depletion of the subsurface magma chamber (from the Latin *calderia* or 'boiling pot').

Carbonate A mineral that contains the compound CO_3, such as calcium carbonate ($CaCO_3$). Thus, carbonate rocks (limestone) are primarily composed of carbonate minerals, skeletal material (corals, flint) or organisms (shells).

Carbonatite A rare type of igneous rock dominated by carbonates, rather than silicates, rich in calcium and containing anomalously high amounts of potassium and/or sodium.

Chert A hard, microcrystalline rock comprised almost entirely of silica (SiO_2). A class of sedimentary rocks ascribed to chemical precipitation.

Cichlid An extensive group of fishes in which new species are continuously being discovered; one of the most extensive family of vertebrates. Includes the commercially important variety, **tilapia**, and important in studies of speciation in evolution due to their great diversity, with estimates of between 2,000 and 3,000 species. Typically found in freshwater but they can also thrive in hypersaline and alkaline lakes.

Crater A general term applied to any bowl-shaped or rounded depression with steep inner slopes, including impact features. A volcanic crater is usually located at the summit of a volcanic cone and is formed by either collapse induced by an explosive eruption or by the accumulation of pyroclastic material onto the rim.

Craton The oldest, most stable component of continental crust that has remained undeformed for several billions of years. Cratons are far more deep-rooted than the younger components of the crust such as mobile belts.

Crust The outermost layer or shell of the Earth defined by factors such as seismic velocity, density and composition and located above the Mohorovičić discontinuity. **Continental crust** is discriminated from **oceanic crust** by being considerably thicker (35–60 km; as compared to 5–10 km for oceanic crust), less dense and compositionally far more heterogeneous. Continental crust is dominated by granitic and sedimentary rocks; oceanic crust by basaltic rocks.

Debris Avalanche Deposit (DAD) Deposits of unsorted, mostly volcanic debris (loose material) that result from the partial or total collapse of volcanic cones; debris is transported by avalanches entirely independent of water.

Deformation A general term for the processes of folding, faulting, shearing, compression or extension of rocks as a consequence of forces associated with a dynamic Earth.

Detrital Pertaining to minerals or rocks formed from **detritus** (loose material), typically transported on surface and deposited in sedimentary basins.

Dome An uplifted or anticlinal structure, circular or elliptical in outline, in which the rocks dip away from the centre in all directions (geology) or a generalised description of an elevated region of plateaus rather than fold mountains (geomorphology).

Dyke An igneous intrusion, tabular in shape, that cuts across the bedding of the country rocks, and typically relatively small and shallow. Most dykes are located vertical or near vertical and vary in thickness from a few centimetres to hundreds of metres.

East African Rift System (EARS) One of the most iconic geological features on Earth, including regional faulting, volcanism, doming and formation of small sedimentary basins within essentially linear structures that extend from Ethiopia to Mozambique. Related to the rifting, or pulling part, of the African plate.

Erosion Surface A land surface, typically near-level, shaped and rounded by the action of erosion and which correlates with a specific period, or age.

Escarpment A linear, more or less continuous steeply inclined slope, locally a cliff, facing in one direction that extends for several or many tens of kilometres and formed by erosion and/or faulting. In the context of the EARS, escarpments invariably represent the surface expressions of major boundary faults.

Extrusive Igneous rocks that have been erupted onto the Earth's surface (synonym: volcanic).

Fault A fracture or a zone of fractures along which there has been significant displacement of the rock sequences on either side.

Fault Breccia The filling of a fault zone by angular fragments derived from the crushing and shattering of the adjacent rock sequences.

Feldspar The most abundant group of rock-forming silicate minerals comprised of two broad groups, plagioclase (calcium–iron–aluminium silicates) and alkali feldspar (sodium–potassium–aluminium silicates). They are typically leucocratic and are the dominant component of granites; also found in many basalts.

Feldspathoid A group of rare rock-forming minerals that are too poor in silica to form feldspar, and include sodium and potassium members, nepheline and leucite, respectively. They can never be found with quartz and are restricted to undersaturated alkaline igneous rocks.

Felsic A generalised term for light-coloured minerals (feldspar, quartz) or light-coloured igneous rocks.

Flood Basalt Eruptions of relatively fluid basaltic lavas, typically from multiple fissures rather than discrete conduits, on a regional scale that smooth out irregularities in older land surfaces.

Foidite A general term for igneous rocks dominated by feldspathoid minerals.

Fractionation The separation of chemical elements by natural processes such as preferential concentration into a magma (partial melting), specific minerals (fractional crystallisation of a magma) or differential solubility (rock weathering).

Fumarole A small vent of volcanic origin from which gases or vapours are emitted, typically associated with geothermal fields in volcanic systems that are either dormant or waning.

Geothermal The internal heat of the Earth related to the breakdown of radioactive elements, an essential component of sustaining plate movements and possibly all forms of life.

Geothermal Energy Heat that can be extracted from the Earth. Most areas of localised high heat flow in the crust are ascribed to volcanism and to a lesser extent radioactive decay of large plutons.

Geothermal Features Related to heat associated with magma chambers and active or dormant volcanoes, e.g. fumaroles, hot springs.

Geyser A type of hot spring that intermittently erupts jets of hot water and steam, derived from geothermal heating of groundwater which periodically converts to steam in a restricted space, prior to being forced to surface through a series of fractures. A group of geysers can be described as a **geyser basin** and the deposits or mounds at the bases of geysers consist of **geyserite**.

Glacial Epoch A period in the Earth's history when extensive ice sheets and glaciers occur on parts of the continental landmass.

Glacial Period An interval during an **Ice Age** when temperatures are generally coldest and ice sheets and glaciers are at their maximum. The coldest peak time is described as a **Glacial Maximum**.

Gneiss A foliated metamorphic rock consisting of bands of granular minerals (often leucocratic feldspars and quartz) that alternate with bands of flaky or platy minerals (often melanocratic mica and hornblende). The most common form of gneiss has a granitic composition and is referred to as granite gneiss.

Gondwana The ancient supercontinent of the Southern Hemisphere that assembled in the Late Palaeozoic and named after the Gondwana system of India (Carboniferous to Jurassic), which in turn is named after the Gonds, the oldest inhabitants of the Indian subcontinent.

Graben An elongate, depressed unit of the crust bounded by faults on the long sides. **Half-grabens** are depressed units in which boundary faults are developed on one side only. The geomorphologic expression of a graben is a rift valley, of a half-graben is a step or inclined valley and of a boundary fault is an escarpment.

Granite A granular, quartz-bearing plutonic rock dominated by leucocratic minerals (feldspar and quartz), with subordinate amounts of mica and hornblende which typically occurs in large bodies (plutons). The dominant constituent of continental crust.

Granulite A metamorphic rock consisting of approximately even sized, interlocking mineral grains with no preferred orientation or fabric.

Granulite Facies The grade or metamorphic facies reached in rocks subjected to deep-seated thermal metamorphism at temperatures >650 °C. The mineralogy is dependent on the primary lithology; schist, for example, may include aluminium silicates such as cordierite, kyanite and sillimanite.

Greenstones A field term applied to dark green, metamorphosed, typically basic igneous rocks characterised by the presence of chlorite or epidote.

Greenstone Belt Elongate terranes dominated by greenstones, typically associated with complex volcano-sedimentary piles of Archaean or Palaeoproterozoic age.

Gregory Rift The Eastern Branch of the EARS, named after geologist and explorer J. W. Gregory.

Geopark A Geopark is a unified area that advances the protection and use of geological heritage in a sustainable way, and promotes the economic well-being of the people who live there. There are Global Geoparks and National Geoparks (*Wikipedia*).

Hard Pan A general term for a hard, impervious, and often clay-rich layer located at or near surface produced from cementation of particles in a soil by precipitation, i.e. of silica (forming silcrete) or calcium carbonate (forming calcrete), generally from seasonal movement of groundwater.

Hominin A more refined group of primates that includes modern and extinct humans (*Homo sapiens*), all of our immediate ancestors, the genus *Hominina* (with its three branches *Homo erectus or Homo Ergaster; Homo heidelbergenesis; Homo habilis*) and the genera *Australopithecine* (including *Australopithecus africanus* and *Australopithecus afarensis*) and *Panina* (chimpanzees). This definition is preferred by zoologists and geneticists as the above species probably share >99% of their DNA. Anthropologists and others may restrict the definition to only *Homo sapiens* and our direct ancestors (*Hominina*).

Hominid A broad group that includes all modern Great Apes and their extinct ancestors (e.g. *Proconsul africanus*) as well as humans and their ancestors.

Homo sapiens The scientific name for the only extant human species. **Homo** is the human genus which includes extinct species of hominin. **Sapiens** (or wise) is the only surviving species of this genus.

Horst An elongate, uplifted unit of the crust bounded by faults on its long sides which is commonly observed as a geomorphological feature i.e. a rift platform.

Hydrothermal Pertaining to hot water and deposits formed from hot brines.

Hypabyssal An igneous rock intruded at shallow depths between deep-seated plutons and surface volcanism, and typically found in either dykes or sills.

Ice Age A loosely used synonym of glacial epoch, a period of extensive glacial activity.

Ice Cores The material recovered from drilling of a glacier or ice sheet which may provide remarkably detailed information on palaeo-climates.

Igneous A rock or mineral that crystallised from molten or partially molten material (magma). Igneous rocks are one of the three main categories of rocks (cf. metamorphic and sedimentary) and can be subdivided into plutonic, hypabyssal and volcanic.

Ignimbrite A volcanic rock of lapilli-rich tuff resulting from violent eruptions, i.e. pyroclastic flows, often associated with formation of calderas.

Intrusive Igneous rocks that occur within pre-existing rock sequences within the Earth's interior, i.e. derived from magma that solidifies prior to reaching the Earth's surface.

Isotope Various forms of an element which has the same number of protons in their nucleus but differing numbers of neutrons (and thus differing atomic weight). A well-known example is uranium which can occur with differing atomic masses including U^{235} and U^{238}.

Kimberlite A relatively rare type of intrusive alkaline igneous rock which is the main source of diamonds.

Lahar Superficial volcanic deposits mobilised by water, including mudflows and debris with primary lava blocks and epiclastic material, i.e. cemented fragments from earlier formed rock sequences.

Leucocratic Applied to light-coloured igneous rocks, typically with >65–70% felsic minerals.

Limestone A type of sedimentary rock dominated by calcium carbonite that commonly forms as a chemical precipitate in warm, shallow water or from a concentration of shell and coral fragments.

Lithosphere The solid portion of the Earth comprised of the Crust and part of the Upper Mantle, typically to a combined depth of approximately 100 km. The Lithosphere is distinguished from the underlying **asthenosphere** by being relatively rigid and less readily deformed, i.e. in plate tectonics theory, the rigid lithospheric plates ride upon the weaker, elastic asthenosphere.

Lineation A generalised term for any linear structure in either a rock or a geological terrane.

Mafic A generalised term for dark-coloured, ferromagnesian minerals (pyroxene, olivine, amphibole) or moderately dark-coloured igneous rock.

Magma Naturally occurring molten rock material generated within the Earth, characterised by being highly mobile, which may be transported upwards into the shallow crust to from either intrusive or extrusive igneous rocks.

Magma Chamber A naturally occurring reservoir of magma, typically located in the crust at a depth of a few kilometres to tens of kilometres formed from molten material which ascended from much deeper in the lithosphere. Many volcanoes are fed from magma chambers within which a complex array of magmas may be fractionated.

Magma or Mantle Plume An upwelling of abnormally hot rock derived in the asthenosphere and which may extend as buoyant columns creating huge volumes of magma by the process of decompression melting in the lithosphere. They feed giant volcanic provinces such as that associated with the EARS.

Mantle The annular zone of the Earth located between the crust and the core, hot yet solid and dominated by iron and magnesian silicates and oxides. Part of the Upper Mantle is incorporated with the Crust into the Lithosphere as it includes a disconformity as significant as the Mohorovičić discontinuity.

Marble A granular rock consisting primarily of recrystallised carbonates (calcite and/or dolomite) which has formed from metamorphism of limestone or dolomite.

Melanistic Dark-coloured morphs of some species of animals, e.g. leopards known colloquially as black panthers.

Melanocratic Applied to dark-coloured igneous rocks, typically with >65–70% mafic minerals.

Mesocratic Applied to igneous rocks, typically with approximately equal proportions of felsic and mafic minerals.

Metamorphic Pertaining to the process of **metamorphism** (changing or altering the primary minerals) and describing rocks that form due to combinations of pressure, heat, fluid flow and shearing stress within the Earth's crust, such as gneiss, schist and marble. Metamorphic rocks may be further classified as meta-igneous or meta-sediments, dependent on their primary origin.

Mica A group of silicate minerals which have a sheet-like or platy form. Found as a primary component of granite and some volcanic rocks, as well as in metamorphic rocks such as schist and gneiss.

Microcontinent A relatively small continental landmass, typically located to a discrete **microplate** (i.e. a very small tectonic plate) that may assemble at a later stage to form a large continent.

Milankovitch Cycle A description of cyclic variations of climate that respond to eccentricities of the Earth's orbital geometry, including angle of tilt of the axis, possibly in increments of 100,000, 43,000, 24,000 and 19,000 years.

Mobile Belt A long, relatively narrow crustal terrane of tectonic activity on a regional scale.

Natrocarbonatite A sodium-rich type of carbonatite, i.e. a volcanic rock that consists mainly of carbonate minerals (rather than silicates) and has crystallised from a molten carbonate magma (rather than a silicate magma).

Natron A naturally occurring mineral salt, sodium carbonate decahydrate ($Na_2CO_3.10H_2O$), which is very soluble in water and which in nature is typically mixed with sodium bicarbonate ($NaHCO_3$).

Nephelinite A fine-grained volcanic or hypabyssal rock of silica-undersaturated basaltic character primarily composed of **nepheline** (a sodic feldspathoid) and pyroxene and lacking olivine and feldspar.

Nepheline Syenite A plutonic rock primarily composed of alkali feldspar and nepheline which crystallises from alkali-rich basaltic magmas (foidite).

Obsidian A black or dark-coloured volcanic glass, usually of rhyolite composition, i.e. highly siliceous, and characterised by conchoidal fractures. Used by ancient cultures for stone tools.

Olivine A dark, ferromagnesian silicate mineral found in mafic igneous rocks.

Orogeny Episodes of mountain building, often resulting from collision of continental plates which causes chains or belts of mountains to form.

Palaeo A prefix denoting great age or remoteness in terms of time, including prehistoric.

Pangea A loosely assembled supercontinent that existed at approximately 250–200 Ma and including almost all of the Earth's continents and microcontinents.

Partial Melting A process by which the relatively low-temperature components of the crust or upper mantle melt to generate a magma, leaving behind a refractory component.

Phonolite A fine-grained extrusive basaltic rock primarily composed of alkali feldspar, and with nepheline as the main feldspathoid.

Phreatomagmatic An explosive style of volcanic eruption that extrudes both magmatic gases and steam, typically caused by the interaction of magma with groundwater or ice.

Plate Any of the mobile parts into which the lithosphere is fractured.

Plateau A term used in geomorphology to describe a nearly flat land surface of great extent and at a relatively high altitude, i.e. synonymous with an elevated plain.

Plate Tectonics The study of the mechanisms by which plates can separate, collide, shear (slide sideways by means of transform faults), subduct (slide beneath each other) or obduct (thrust up over one another) through geological time.

Plinian-style eruption An explosive style of volcanism named after Pliny the younger who documented the catastrophic AD79 eruption of Vesuvius and characterised by sustained eruptions that include high eruptive columns that generate copious amounts of pumice. They may represent the ejection of thousands of cubic kilometres of magma.

Plutonic An igneous rock formed at great depth, thus a **pluton** is a large intrusive body of coarsely crystalline rocks which cooled slowly. Although plutons originally form at depth, they can crop out on surface due to erosion stripping away the surface cover.

Primate A group of mammals that includes lemurs, monkeys, apes and humans.

Proto A prefix denoting first, or earliest form of.

Pumice A volcanic rock, typically glass foam with abundant vesicles (cavities), often sufficiently buoyant to float on water, and generally ejected from vents. The composition can be siliceous (rhyolite) or basaltic.

Pyroclastic Volcanic material formed by the fragmentation of magma and/or pre-existing rock (including conduit walls), typically ejected from a vent by explosive activity, and containing particles of chilled melt, (including pumice), earlier formed crystals (phenocrysts) and lithic material. Products include ash particles, **lapilli** (small fragments), volcanic bombs, breccia and agglomerate. The amount of vesicular material in the **pyroclasts** determines the formation of products that include pumice, scoria or cinder and spatter.

Pyroxene A dark, ferromagnesian silicate mineral found in mafic igneous rocks.

Quartz A common mineral comprised of crystalline silica (SiO_2).

Quartzite A rock with a granular texture dominated by quartz and formed by recrystallisation of sandstone by regional or thermal metamorphism.

Radiometric The measurement of geologic time by the study of isotopic abundances of specific radioactive elements that disintegrate at a known rate.

Rhyolite An extrusive igneous rock that commonly exhibits flow textures and is characterised by highly siliceous compositions. In comparison to basalts, rhyolite magma is notably viscous and much richer in volatiles, often resulting in explosive eruptions.

Rifting The separation of geological terranes, including the drifting or pulling-apart of continents during formation of oceanic basins. In continental separation, rifting eventually results in formation of oceans and seafloor spreading centres.

Rift Platform A geomorphological description of the uplifted blocks on either side of a rift valley.

Rift Valley A geomorphological description of a linear, depressed block.

Rodinia Interpreted as the first and oldest supercontinent with an age of approximately 1.100 Ga.

Salt A general description of a compound that is produced from a chemical reaction between an acid and a base and in which the acid's hydrogen atoms are replaced by metal atoms of the base. Most salts are crystalline ionic compounds, such as sodium chloride (NaCl), or common table salt. Other naturally occurring salts include soda, natron and trona.

Sandstone A sedimentary rock primarily comprised of sand-sized grains of quartz and lesser amounts of feldspar, typically the product of erosion of detrital sediments and deposited by either water or wind.

Saturated A chemical classification of igneous rocks that are dominated by silica and contain common rock-forming minerals such as quartz and feldspar.

Savannah An area dominated by grassland with scattered trees (typically acacia).

Schist A strongly foliated crystalline rock formed by dynamic metamorphism that splits into thin flakes or slabs.

Scoria Pyroclasts or small drops of vesicular magma ejected into the air and deposited as irregular fragments, often forming poorly consolidated cones, near a volcanic vent. Synonymous with cinder.

Sedimentary One of the three primary groups of rocks describing compacted or lithified sediment, i.e. solid, fragmented material that originates from weathering of older rocks and is transported or deposited by air, water or ice, or that accumulates by other natural agents such as chemical precipitation from solution or accretion by organisms. Sedimentary rocks form from layers of unconsolidated material on the Earth's surface at near-ambient temperatures.

Seismic Energy released during an earthquake or similar vibration of the Earth.

Silicate An extensive group of minerals composed of varying proportions of silicon (Si) and oxygen (O) which are typically bonded with elements that include aluminium (Al), magnesian (Mg), iron (Fe), calcium (Ca), sodium (Na) and potassium (K). Silicates make up most of the crust and mantle.

Sill An igneous intrusion, tabular or an elongate, layer emplaced parallel with the country rocks, typically relatively small and shallow.

Soda A commercial term for sodium carbonate (Na_2CO_3) and/or sodium decahydrate ($Na_2CO_3.10H_2O$), naturally occurring salts also known as washing soda.

Soda Ash A crystalline substance, sodium bicarbonate ($NaH.CO_3$) commonly known as baking soda, occurs in nature rarely as nahcolite, and produced from natural salt deposits, typically comprised of natron and trona.

Soda Lake A loosely used term to describe an alkali lake whose waters contain a high content of various sodium salts.

Spatter Blebs of lava ejected close to the vent that weld together on impact, typically forming small cones.

Species Any group of organisms that can interbreed and produce fertile offspring.

Strata Compacted layers or tabular bodies of sedimentary rocks, further classified as a stratigraphic unit, implying a chronological sequence.

Stratigraphy The science of rock strata including an interpretation of the original succession and chronology, typically using fossil assemblages. In recent times, geochronology or radiometric dating enables igneous rock units, i.e. lavas and ashes to be incorporated into a stratigraphic sequence.

Strombolian A style of volcanism characterised by periodic, relatively small-scale eruptions of basaltic ash, lapilli and scoria, typically within or close to the crater.

Structure The general disposition, attitude, arrangement or relative position of rock masses or sequences of a region.

Supercontinent An amalgamation of large and small continental plates into a gigantic continental mass, such as Pangea.

Syncline A downfolded structure in which the youngest rocks occur in the core and the oldest rocks in the outer part.

TAS A chemical plot of total alkalis versus silica used to discriminate the basaltic group of igneous rocks.

Tectonics A branch of geology dealing with the broad architecture of the outer part of the Earth, specifically detailing the structural or deformational features, and including the historical evolution of such features, including **tectonic plates**.

Tephra A general term for material associated with pyroclastic deposits and volcanic ash.

Terrane The description of a specific geologic region or area, often bounded by faults or unconformities and with a different geologic history to adjacent regions.

Trachyte A group of fine-grained, generally porphyritic, extrusive igneous rocks (volcanic) having alkali feldspar and minor mafic minerals as the main constituents.

Triple Junction A point where three lithospheric plates join.

Tuff Rock formed from fine-grained, wind-borne or water-laid volcanic ash.

Trona A naturally occurring mineral salt, sodium sesquicarbonate dihydrate ($Na_3CO_3.HCO_3.2H_2O$) which is generally found mixed with natron.

Unconformity A major break or time gap in the rock record and where rocks of a given time are absent between the underlying (typically older) and overlying (typically younger) sequences. Related to either non-formation or erosion having removed the missing strata.

Undersaturated A chemical classification of igneous rocks that are relatively poor in silica and contain silica-poor minerals, particularly feldspathoids (such as nepheline) and in which silica-rich minerals, such as quartz and feldspar are absent.

Uplift The process by which a structurally high area in the crust is produced by tectonic movements that raise the rock sequence (as in a dome or plateau).

Volcanic Rock types or activity associated with a volcano. Volcanic rocks are igneous rocks extruded onto the surface of the Earth, e.g. as lavas or ash fall.

Volcaniclastic A generalised description of all types of volcanic particles or rocks regardless of their origin, i.e. not only those from a vent (pyroclastic) but also rock fragments released by erosion of earlier formed volcanic material.

Vulcanian Pertaining to volcanic activity similar to that of the active volcano on the island of Vulcano, one of the Aeolian Islands located off the southwest coast of Italy, and characterised by the explosive ejection of lithic fragments (pre-existing rock blocking the conduit as well as new lava and volcanic bombs) and irregular, cannon-like eruptions.

Warp A slight flexure or bend (upwards or downwards) in the Earth's crust, usually on a regional scale.

Xenolith A foreign inclusion in an igneous rock, typically derived from the earlier formed country rock, for example, the sidewalls of a conduit or pluton.

Note Explanations of geological eras, periods, and epochs are excluded as they are defined in Fig. 2.2.

References

Bates, R. L., & Jackson, J. A. (1980). *Glossary of geology* (751 p). Virginia: American Geological Institute.
Redfern, R. (2002). *Origins. The evolution of continents, oceans, and life* (360 p). London: Wiedenfeld and Nicolson.

Index

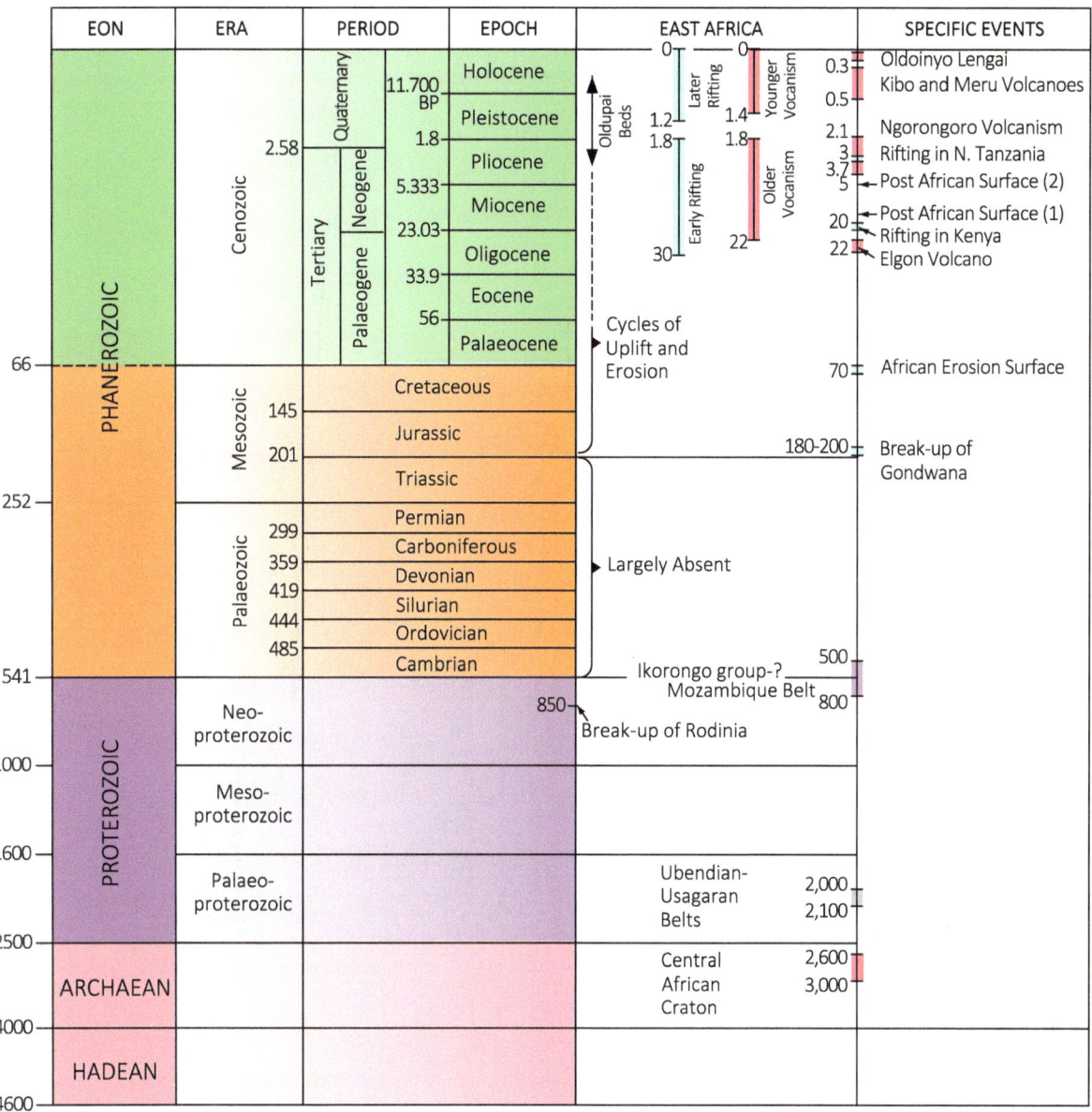

EON	ERA	PERIOD		EPOCH		EAST AFRICA		SPECIFIC EVENTS

Simplified stratigraphic column for East Africa. Ages (Ma) incorporate latest recommendations of the International Union of Geological Sciences with the exception of the Pliocene–Pleistocene boundary which is retained at the traditional age of 1.8 Ma (not 2.58 Ma) as this would impact severely on published literature pertaining to many of the sites in northern Tanzania, including Oldupai Gorge